Alexander Kromp

Model-based Interpretation of the Performance and Degradation of Reformate Fueled Solid Oxide Fuel Cells

Schriften des Instituts für Werkstoffe der Elektrotechnik,
Karlsruher Institut für Technologie

Band 24

Eine Übersicht über alle bisher in dieser Schriftenreihe
erschienene Bände finden Sie am Ende des Buchs.

Model-based Interpretation of the Performance and Degradation of Reformate Fueled Solid Oxide Fuel Cells

by
Alexander Kromp

Dissertation, Karlsruher Institut für Technologie
Fakultät für Elektrotechnik und Informationstechnik, 2013

Impressum

Karlsruher Institut für Technologie (KIT)
KIT Scientific Publishing
Straße am Forum 2
D-76131 Karlsruhe
www.ksp.kit.edu

KIT – Universität des Landes Baden-Württemberg und
nationales Forschungszentrum in der Helmholtz-Gemeinschaft

KIT Scientific Publishing 2013
Print on Demand

ISSN 1868-1603
ISBN 978-3-7315-0006-3

Model-based Interpretation of the Performance and Degradation of Reformate Fueled Solid Oxide Fuel Cells

Zur Erlangung des akademischen Grades eines

DOKTOR-INGENIEURS

der Fakultät für

Elektrotechnik und Informationstechnik

des Karlsruher Instituts für Technologie (KIT)

vorgelegte

DISSERTATION

von

Dipl.-Ing. Alexander Wilhelm Kromp

geb. in Würzburg

Tag der mündlichen Prüfung: 31.01.2013

Hauptreferentin: Prof. Dr.-Ing. Ellen Ivers-Tiffée

Korreferenten: Prof. Robert J. Kee

Prof. Dr. Olaf Deutschmann

众里寻她千百度

for Huilin

Acknowledgements

The present dissertation has been developed during my research work at Institut für Werkstoffe der Elektrotechnik (IWE). First and foremost, I would like to express my gratitude to Prof. Dr.-Ing. Ellen Ivers-Tiffée, who created the framework for this research. I am deeply grateful for the trust she has placed in me, for providing an excellent working environment and for her constant promotion of my professional and personal development.

In addition to the advisement that I received from Prof. Dr.-Ing. Ellen Ivers-Tiffée, I have been fortunate to have two of my mentors act as co-advisors for this thesis work. My first co-advisor, Prof. Robert J. Kee was kind enough to welcome me at the Colorado School of Mines (CSM), where I benefitted strongly from the numerous pleasant and fruitful discussions with him and Prof. Huayang Zhu. I am also indebted to my second co-advisor, Prof. Olaf Deutschman from the Institute for Chemical Technology and Polymer Chemistry (ITCP) at the KIT for his constant interest in my work and for taking the time to mentor me in the Helmholtz Research School Energy-Related Catalysis.

As a member of the Helmholtz Research School, I have enjoyed the education and the inspiring atmosphere, which arose from the diverse scientific background of my schoolmates. I would like to thank the coordinator, Dr. Steffen Tischer for his support in the research school that has helped to realize my research stay in Colorado. I gratefully acknowledge Prof. Neal Sullivan and his research group for welcoming me into the Colorado Fuel Cell Center at CSM.

The foundation of this work has been laid by my diploma thesis advisor and colleague, Dr.-Ing. André Leonide. He has always been a great teacher to me and this work would never have been possible without him. I would also like to thank my group leader, Dr.-Ing. André Weber for constantly sharing his expertise and knowledge during the last three and a half years.

During my time at KIT, I have had the pleasure of working with several students of my own. I deeply appreciate their confidence, their dedication and the excellent collaboration we have had. I would like to emphasize my gratitude to Dipl.-Ing. Helge Geisler and B.Sc. Sebastian Dierickx, whose extraordinary commitment has contributed significantly to this work. I thank all my students for giving me the honor of taking part in their professional development.

I would also like to thank all of the members of the IWE for their steady support and for providing a pleasant working atmosphere. I am particularly indebted to my colleagues Dr.-Ing. Michael Kornely and Dipl.-Math. Jochen Joos for their moral support, which has helped me get through some tough times. I would further like to thank my colleague Christian Gabi for being available 24/7 to support me in all technical matters.

Special thanks go to Dipl.-Chem. Claudia Diehm (ITCP) and Dr.-Ing. Dino Klotz for proofreading this manuscript.

Last but not least I want to thank my family, with my wife, my parents, my sisters and my grandparents leading the way. Your love and confidence has always sustained me and has from time to time put some daylight between myself and my work.

ALEXANDER KROMP

Karlsruhe-Durlach
February 2013

Zusammenfassung

Der Bedarf an nachhaltiger, sauberer und sicherer und Energieerzeugung hat im letzten Jahrzehnt ein großes öffentliches Interesse an der Brennstoffzellentechnologie geweckt. Brennstoffzellen sind in der Lage, chemische Energie, wie sie in Brennstoffen vorliegt, auf direktem Wege in elektrische Energie umzuwandeln. Mit der sauberen und hocheffizienten Umwandlung von Wasserstoff in elektrische Energie haben Brennstoffzellen das Potenzial, eine Schlüsselrolle in der zukünftigen Energieversorgung einzunehmen. Die Hochtemperaturbrennstoffzelle SOFC (engl., Solid Oxide Fuel Cell) ist dabei von besonderem Interesse. Durch die hohen Betriebstemperaturen und die guten katalytischen Eigenschaften der Brenngaselektrode (Anode) können konventionelle kohlenwasserstoffhaltige Brennstoffe direkt umgesetzt werden. Somit ist die SOFC im Gegensatz zu anderen Brennstoffzellensystemen nicht auf Wasserstoff als Brenngas angewiesen und der Einsatz der SOFC ist nicht vom flächendeckenden Ausbau der Wasserstoffinfrastruktur abhängig. Durch die flexible Wahl der Brennstoffe ergibt sich ein breites Spektrum an möglichen Anwendungen von stationären Kombi-Kraftwerken auf Basis von Erdgas zur Produktion von Strom und Wärme für Haushalte über auxiliary power units (Hilfskrafterzeuger, APU) in mobilen Anwendungen, betrieben mit flüssigen Brennstoffen bis hin zum Einsatz in Großkraftwerken auf Basis von Kohlegas.

Ein vorrangiges Entwicklungsziel in der SOFC-Forschung ist neben Langzeitstabilität und Kostenreduktion auch die Leistungsfähigkeit der Zellen. Die Leistungsfähigkeit von Einzelzellen beruht im Wesentlichen auf den elektrochemischen Verlustprozessen, die im Betrieb der SOFC auftreten. Für den Betrieb mit kohlenwasserstoffhaltigen Brennstoffen ist bis dato kein umfassendes Verständnis der einzelnen Verlustprozesse vorhanden.

Deshalb war es das vorrangige Ziel dieser Arbeit, die einzelnen Verlustprozesse mit geeigneten Messmethoden zu erfassen, diese zu identifizieren und letztendlich den zugrundeliegenden physikalischen Prozessen zuzuweisen.

Für die Experimente wurde ein anodengestütztes Zellkonzept (engl., anode supported cell, ASC) mit einer Ni/8YSZ-Anode gewählt. Um eine eindeutige Charakterisierung der stark von den Betriebsbedingungen abhängigen Verlustprozesse zu realisieren, wurde die Anode mit einem Modell-Reformat betrieben, was Gasmischungen von H_2, H_2O, CO und CO_2 im jeweiligen chemischen Gleichgewicht an der Anode entspricht.

Im Einzelnen wurden in dieser Arbeit die im Folgenden zusammengefassten, maßgeblichen Ergebnisse erzielt:

Identifikation der anodenseitigen Verlustprozesse im Reformatbetrieb (Kapitel 4)

Mittels elektrochemischer Impedanzspektroskopie wurden die einzelnen Verlustprozesse an reformatbetriebenen ASCs untersucht. Die am Institut für Werkstoffe der Elektrotechnik (IWE) entwickelte DRT-Methode (engl., Distribution of Relaxation Times) ermöglichte die separierte Betrachtung der einzelnen, in den gemessenen Impedanzspektren enthaltenen elektrochemischen Verlustprozesse. Durch die systematische Variation einer gezielten Auswahl an Betriebsparametern konnten vier Polarisationsprozesse an der Anode identifiziert und physikalisch eingeordnet werden. Neben den Prozessen, die der Gasdiffusion im Anodensubstrat (P_{1A}) und der Kopplung von der Ladungstransferreaktion und Ionentransport in der Anodenfunktionsschicht (P_{2A}, P_{3A}) zugeordnet werden konnten, weisen die Spektren einen weiteren elektrochemischen Prozess bei charakteristischen Frequenzen unterhalb von 1 Hz auf. Dieser Prozess tritt ausschließlich im Reformatbetrieb auf und wird daher im Folgenden als P_{ref} bezeichnet. Die untersuchten Parameterabhängigkeiten lassen auf eine Zuweisung zu Gasdiffusion, evtl. gekoppelt mit den Reformierungsreaktionen, schließen.

Anhand dieser Ergebnisse war es möglich, ein physikalisch motiviertes Ersatzschaltbildmodell zu entwickeln, welches die Impedanzantwort von reformatbetriebenen ASCs wiedergibt. Jeder der identifizierten elektrochemischen Verlustprozesse an der realen Zelle wird im Ersatzschaltbild durch ein passendes Ersatzschaltbildelement dargestellt. Durch den Fit der simulierten an gemessene Impedanzspektren können die einzelnen Verlustprozesse getrennt voneinander quantitativ analysiert werden.

Kinetische Untersuchung der Elektrochemischen Brennstoffoxidation an der Anode (Kapitel 5)

In einer reaktionskinetischen Analyse wurde der Mechanismus der elektrochemischen Brenngasoxidation an der Anode bestimmt. Hierzu wurden die Aktivierungspolarisationsprozesse der Anode (P_{2A} und P_{3A}), deren maßgeblicher Bestandteil die Landungstransferreaktion ist, für den Betrieb mit Wasserstoff (Brenngasmischungen aus H_2 und H_2O) und Reformatgasen verglichen. Für die getrennt durchgeführten Variationen der H_2- und der H_2O-Konzentrationen an der Anode, sowie für die Variation der Betriebstemperatur konnte das gleiche qualitative Verhalten der Prozesse in beiden Betriebsmodi festgestellt werden. Durch Bestimmung der Aktivierungspolarisationswiderstände ($R_{act,an}$ = R_{2A} + R_{3A}) für die durchgeführten Parametervariationen mittels Ersatzschaltbild-Fit konnten die kinetischen Parameter der elektrochemischen Brennstoffoxidation ermittelt werden. Alle drei Parameter (Reaktionsordnungen im Bezug auf H_2 und H_2O sowie Aktivierungsenergie) zeigten Übereinstimmung für den Betrieb mit Wasserstoff und mit Reformatgasen. Damit konnte eindeutig gezeigt werden, dass für den Reformatbetrieb nur H_2 elektrochemisch oxidiert wird. Dies bedeutet im Weiteren, dass die übrigen, prinzipiell elektrochemisch aktiven Spezies im Brenngas über die katalytisch aktivierten Reformierungsreaktionen im Anodensubstrat thermisch oxidiert werden. Im Falle von CO dürfte der Reaktionsschritt dieser vorgelagerten Oxidation über die Wassergas-Shift (WGS) Reaktion ablaufen.

Verständnis der Gastransportprozesse im reformathaltigen Brenngas der Anode (Kapitel 6)

Die anodenseitigen Stofftransportprozesse wurden mit Hilfe von Impedanzspektroskopie und der Finiten Elemente Methode (FEM) untersucht. Grundsätzlich müssen die Brennstoffe der elektrochemischen Anodenreaktion der Anode zugeführt und die Produkte abtransportiert werden, was im porösen Anodensubstrat mittels Gasdiffusion geschieht. Im Falle der reformatbetriebenen SOFC-Anode ergeben sich mit den Erkenntnissen aus dem vorangegangenen Kapitel zwei Transportpfade (H_2/H_2O und CO/CO_2), die durch die WGS-Reaktion miteinander verknüpft sind. Die Implementierung dieses schematischen Verständnisses in FEM-Simulation hat gezeigt, dass die Impedanzantwort des gekoppelten Netzwerks von Diffusion und Reaktion die gemessene niederfrequente Impedanzantwort aus den Prozessen P_{1A} und P_{ref} wiedergibt. Durch die anschließende Untersuchung der zugrundeliegenden Stofftransportprozesse mittels FEM-Simulation konnten die gemessenen Polarisationsprozesse ihrem physikalischen Hintergrund zugewiesen werden. Für den schnelleren Prozess P_{1A} wurde gezeigt, dass dieser von der schnellen Diffusion von H_2/H_2O verursacht wird. Beim niederfrequenteren Prozess P_{ref} hingegen wurde ein deutlicher Einfluss der Diffusion von CO/CO_2 sichtbar.

Einfluss von Schwefel auf die anodenseitigen Verlustprozesse (Kapitel 7)

Abschließend wurden die in dieser Arbeit gewonnenen Erkenntnisse genutzt, um den Einfluss von Schwefel auf die Leistung von SOFC-Anoden zu untersuchen. Viele Brennstoffe enthalten Spuren des als Katalysatorgift bekannten Stoffes, welcher die Leistung von SOFC-Anoden stark beeinträchtigt. Dabei wurde zum ersten Mal in der Literatur eine getrennte Betrachtung der einzelnen elektrochemischen Verlustprozesse durchgeführt. Im Betrieb mit schwefelhaltigen Reformatgasen wurde eine starke Degradation der anodenseitigen Verlustprozesse festgestellt. Am stärksten betroffen waren die Aktivierungspolarisationsprozesse P_{2A} und P_{3A}, deren Gesamtwiderstand $R_{act,an}$ um ein Vielfaches anstieg. Es konnte gezeigt werden, dass dieser Anstieg hauptsächlich auf die Auswirkungen der Katalysatorvergiftung durch Schwefel auf die elektrochemische Brennstoffoxidation an der Anode zurückzuführen ist. Auch die zuvor analysierten Gastransportverluste zeigten eine deutliche Degradation durch die Schwefelbeaufschlagung. Während der Prozess P_{1A}, welcher die Diffusion von H_2/H_2O wiedergibt, stark angestiegen ist, ist der Widerstand des von der CO/CO_2-Diffusion dominierten Prozesses (P_{ref}) hingegen fast auf den Wert null abgesunken. Dieses Verhalten konnte auf die Vergiftung der WGS-Reaktion zurückgeführt werden, welche einen stark verminderten Umsatz von CO zur Folge hat. Um den Einfluss von Schwefel auf die Leistung von SOFC-Anoden zu verringern, müssen beide Effekte berücksichtigt werden.

Mit den in dieser Dissertation erarbeiteten Ergebnissen wurde erstmals ein umfassendes Verständnis der elektrochemischen Verlustprozesse an reformatbetriebenen SOFC-Anoden geschaffen. Mittels elektrochemischer Impedanzspektroskopie und geeigneten Auswertemethoden konnten die einzelnen Verlustprozesse eindeutig identifiziert werden und den folgenden physikalischen Vorgängen an der Anode zugewiesen werden:

- Ladungstransferreaktion — Elektrochemische Oxidation von H_2

- Brenngastransport — Diffusion über zwei Transportpfade (H_2/H_2O und CO/CO_2), gekoppelt durch die WGS-Reaktion

Die in dieser Arbeit abschließend durchgeführte Vergiftungsanalyse hat die Gültigkeit der gewonnenen Erkenntnisse bestätigt. Mit der vorliegenden Untersuchung wurde zudem ein Verfahren gezeigt, welches in der Lage ist, den Einfluss der Schwefelvergiftung in SOFC-Anoden auf (i) die elektrochemische Brennstoffoxidation und (ii) die Reformierungsreaktionen getrennt voneinander zu untersuchen.

Contents

1. Introduction

Electrochemical energy converting devices are have the potential to play an important role in storage and generation of electrical energy of future energy scenarios. Other than conventional forms of power generation, they convert chemical energy (which is, e.g., stored in a fuel) directly into electrical energy. In fuel cells, spatially separated partial electrochemical reactions avoid the long chain of combustion – thermal energy – mechanical energy – electrical energy. Accordingly, fuel cells can achieve significantly higher efficiencies than conventional power generation, hence promising reduced fuel consumption and CO_2 emissions.

Amongst several fuel cell system, the solid oxide fuel cell (SOFC) has been attaining special attention. In contrast to other fuel cell systems, SOFC technology does not rely on hydrogen as fuel and is hence not depending on breakthroughs in the expansion of the hydrogen infrastructure. The high operating temperatures and the good catalytic activity of the fuel electrodes (anodes) facilitate the operation on a variety of readily available hydrocarbon fuels. Possible fuels range from natural gas to bio-derived fuels to coal-gas. Accordingly, SOFC systems offer a broad range of possible applications from small-scale stationary combined heat and power (CHP) systems for residential use to auxiliary power unit (APU) in mobile applications to large-scale power generation.

Current SOFC development aims at long-term stability and cost reduction. For the SOFC development, which aims at long-term stability and cost reduction, it is essential to have an overall understanding of the electrochemical loss processes that determine the performance of the single cells.

In the most general sense, the origin of losses can be traced back to ionic and electronic transport (ohmic losses), charge transfer reactions at the electrodes and diffusive transport of the gas species involved in the electrode reactions (electrochemical polarization losses). For operation with reformate fuels, the anodic charge transfer reaction, i.e., the electrochemical fuel oxidation at the anode and the multi-component gas transport through the porous anode structure is of high interest.

There is a great deal of knowledge about the conversion of readily available fuels at SOFC anodes in terms of catalytic reforming chemistry, whereas little is known about the electrochemistry of the cells. To this day, there is no overall understanding for operation on readily available fuels available in literature.

Goals of this Work

The aim of this work is to capture and understand the electrochemical loss processes of the SOFC for for operation with hydrocarbon fuels. In particular, this comprises the identification of (i) the electrochemical fuel oxidation mechanism, (ii) the fuel gas transport properties in porous electrode structures and (iii) the interaction with heterogeneous reforming chemistry with the above mentioned processes.

As subject to the investigations, state of the art anode supported single cells manufactured by Forschungszentrum Jülich (FZJ) are chosen. This cell type is one of the best-performing, most stable and hence most investigated SOFC systems available. The anode supported cells (ASCs) were built on cermet anode structures made of nickel (Ni) and yttria stabilized zirconia (YSZ), which is the most common SOFC anode material system.

Electrochemical Impedance Spectroscopy (EIS) is applied in order to measure the electrochemical loss processes on the single cells. Along with the distribution of relaxation times (DRT) method, the individual processes can be identified and assigned to their physical origin. With this knowledge, a physically meaningful equivalent circuit model (ECM) can be set up, representing the impedance response of the analyzed cells under reformate operation. The subsequent complex nonlinear least squares (CNLS) fit procedure of the ECM to the measured spectra allows for a quantitative analysis of the individual loss processes.

The highly endothermic reforming reactions induce significant gradients in temperature and species concentration at the SOFC anodes, thus inhibiting detailed electrochemical analyses. Therefore, in this work the operation on model reformate fuels (gas mixtures of H_2, H_2O, CO, CO_2 and N_2 at chemical equilibrium) is developed. The applied model fuels represent fuel gas mixtures of H_2, H_2O, CO, CO_2 and N_2 at chemical equilibrium. With this method, species conversion via the reforming reactions is inhibited and analyses under strictly defined operating parameters are guaranteed.

After identification of the individual loss processes under reformate operation and the subsequent assignment of their physical origin, the electrochemical fuel oxidation mechanisms will be analyzed. From the CNLS-fit of the ECM kinetic parameters can be deduced and compared to electrochemical oxidation reactions that have already been studied.

Gas transport properties in the porous structure of the reformate fueled anode will be analyzed via a combined EIS and finite element method (FEM) modeling study. It is aimed to identify gas transport pathways and possible interactions with reforming chemistry.

Outline

In the following Chapter 2, an introduction to the fundamentals needed to understand the core of this thesis is given.

Chapter 3 refers to the technical realization of the experiments performed in this work. It contains descriptions of (i) the cell-design and material composition of the analyzed ASC single cells, (ii) the measurement setup used for the electrochemical characterization and (iii) the operating parameter range over which the measurements were carried out, focusing on the application of model reformate fuels.

The subsequent chapters are the main part of this thesis, where the results of this work are presented:

- In Chapter 4, the electrochemical loss processes occurring under reformate operation of the analyzed cells are identified and physically interpreted. As a result from this study, a physically meaningful ECM is set up.

- In Chapter 5, the developed ECM is applied in order to perform a kinetic analysis on the anode activation polarization processes. With this analysis, the electrochemical fuel oxidation mechanism for reformate fueled Ni/YSZ anodes is identified.

- In Chapter 6, the gas transport properties are analyzed via a combined EIS and FEM modeling study.

- In Chapter 7, the methods and findings presented in the preceding chapters are applied in order to investigate the impact of sulfur-poisoning on the individual electrochemical loss processes of the analyzed cells.

A brief summary of the essential findings of this work is given in the last Chapter 9.

2. Fundamentals

This chapter introduces the reader to the fundamentals, upon which this thesis is built. The explanations on working principle and material systems of solid oxide fuel cells (SOFCs) are deliberately brief. Interested readers are referred to the relevant handbooks [1–3]. More attention is addressed to the present understanding of the electrochemical loss mechanisms of SOFCs and their measurablity via electrochemical measuring techniques. Both topics are essential for the understanding of the present thesis. As a last point, the reader is introduced to the topic of catalytic reforming chemistry in SOFC anodes.

2.1. Working Principle of Solid Oxide Fuel Cells

Fuel cells are, along with accumulators and batteries, electrochemical energy converters based on the principle of galvanic cells. These devices are capable to convert stored chemical energy directly into electrical energy. In fuel cells, the chemical energy from a fuel is converted into electricity through a chemical reaction with an oxidant. The reactants and the products are continuously supplied and removed. Figure 2.1 illustrates the fuel cell working principle using the example of a SOFC. In the SOFC, the gaseous fuel (e.g., hydrogen, H_2) and oxidant (e.g., oxygen, O_2) are separated by a gas-tight ceramic membrane, inhibiting the direct reaction of fuel and oxidant (combustion) as:

$$H_2 + \tfrac{1}{2} O_2 \rightleftharpoons H_2O. \tag{2.1}$$

At high operating temperatures (above $500\,°C$), the solid oxide ceramic membrane becomes conducting for O^{2-} ions, thus acting as **electrolyte**. Electrodes placed at both sides of the electrolyte facilitate the incorporation of O^{2-} ions into and their removal from the electrolyte. This has the result that the overall fuel oxidation reaction (Eq. 2.1) proceeds via two spatially divided, electrochemical partial reactions.

At the **oxygen electrode (cathode)**, oxygen from the gas phase dissociates, accepts two electrons and is incorporated into the electrolyte phase as:

$$\tfrac{1}{2} O_{2(g)} + 2\,e^- \rightleftharpoons O^{2-}_{(el)}. \tag{2.2}$$

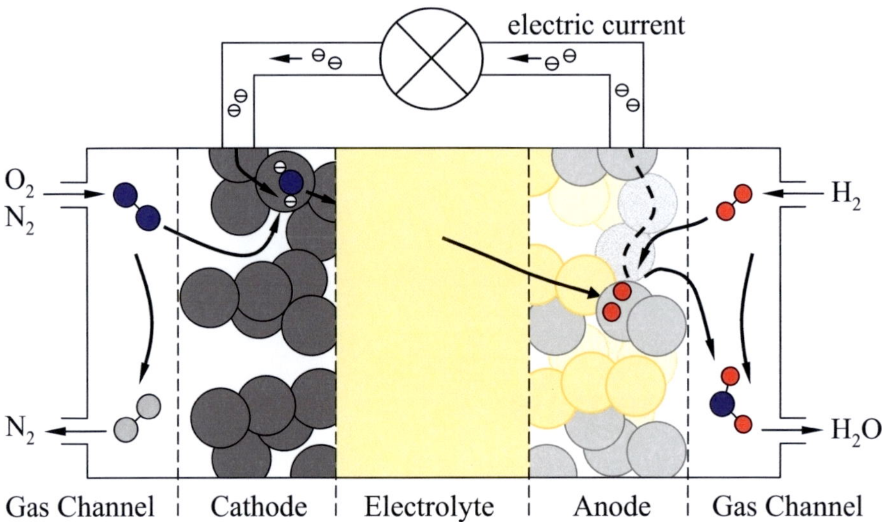

Figure 2.1.: Function principle of the SOFC (adapted from Ref. [4])

Due to the chemical potential drop between fuel and oxidant, O^{2-} ions diffuse through the electrolyte toward the **fuel electrode (anode)**, where they react with H_2 to water (H_2O). In this electrochemical fuel oxidation, two electrons are released:

$$O^{2-}_{(el)} + H_{2(g)} \rightleftharpoons H_2O_{(g)} + 2\,e^-. \tag{2.3}$$

The reversible electrode reactions (Eqs. 2.2 and 2.3) reach chemical equilibria and between anode and cathode, a terminal voltage can be measured. This resulting voltage compensates the chemical potential between fuel and oxidant and can thus be calculated with the Gibbs free energy ΔG of the overall cell reaction (Eq. 2.1):

$$U_{\text{th}} = -\frac{\Delta G}{zF}. \tag{2.4}$$

The theoretical Nernst voltage, or Nernst potential, U_{th}, is named after the German chemist Walther Nernst, who first formulated this relation in 1889 [5]. For the SOFC with H_2 as fuel, the number of electrons transferred per reacting mole of fuel is: $z = 2$. F is the Faraday constant. With the temperature dependence of the Gibbs energy for an ideal gas

$$\Delta G = \Delta G^0 + RT \ln K, \tag{2.5}$$

and the equilibrium constant K of the overall cell reaction (Eq. 2.1) as

$$K = \frac{\sqrt{pO_{2,\text{cat}}} \cdot pH_{2,\text{an}}}{pH_2O_{\text{an}}}, \tag{2.6}$$

a more practical formulation can be deduced for the Nernst voltage of a SOFC. In dependence on the equilibrium partial pressures of the reactants at both electrodes, Eq. 2.4 can hence be rewritten as [6, 7]

$$U_{\mathrm{N}} = -\frac{\Delta G^0}{2F} + \frac{RT}{2F} \ln \left(\frac{\sqrt{p\mathrm{O}_{2,\mathrm{cat}}} \cdot p\mathrm{H}_{2,\mathrm{an}}}{p\mathrm{H}_2\mathrm{O}_{\mathrm{an}}} \right). \tag{2.7}$$

In the typical operation range of a SOFC operated on H_2 as fuel and air as oxidant ($T = 600...950\,°C$, $1\,\%$ H_2O at the anode), Eq. 2.7 yields values of U_{N} between 1.18 and 1.13 V.

2.2. Materials and Design

As depicted in Fig. 2.1, the setup in form of a membrane electrode assembly (MEA) is essential to the operation of the SOFC. For proper operation of the cell, the functionality of the MEA must guarantee for highly efficient proceeding of the electrode reactions (Eqs. 2.2 and 2.3) as well as fast transport of the involved reactants. The latter implicates transport of gaseous species, solid state ions and electrons. The electrodes must hence exhibit high catalytic activity, high electronic and ionic conductivities combined with high porosity. The electrolyte however has to be purely ionic conducting and gastight. Furthermore, addressing the reliability of the SOFC as electroceramic device, each component must exhibit (i) stability at high temperatures and in reducing and/or oxidizing atmospheres, (ii) chemical compatibility with the neighboring components as well as (iii) matching thermal expansion coefficients.

The long history of the SOFC bears a long history of material optimization to meet the complex requirements of the individual cell components. Several comprehensive overviews can be found in literature [8–15]. In the following paragraphs, only a brief introduction to the basic components of the MEA (i.e., electrolyte, anode and cathode) is provided.

Electrolyte

In order to exploit the chemical potential between fuel and oxidant to the highest possible extent, the ceramic electrolyte of an SOFC must be gas-tight and isolating for electron conduction. Furthermore, the electrolyte must have sufficient conductivity to support the ion current without substantial ohmic losses. Oxygen-ion conducting materials are often fluorite-type oxides (e.g., ZrO_2, CeO_2, BiO_2 and ThO_2). Apart from stabilizing the cubic phase of the ceramic throughout a great temperature range, reasonably high ionic conductivities are achieved through blending with dopants. Common dopants are di- or tri-valent rare-earth or alkaline-earth cations (e.g., Ca, Mg, Y, Nd, Sm, Yb and Sc). The stabilization increases the oxygen vacancy concentration of the electrolyte material, thus enhancing its ionic conductivity. Stabilized zirconia, in particular yttria stabilized zirconia (YSZ) has become the prevalent electrolyte material in SOFCs. This material has been extensively studied as electrolyte material [16–21], and with regards to its application in SOFCs [22–28].

The traditionally projected high operating temperatures (ca. $1000\,°C$) were motivated by the strong temperature dependence of the ionic conductivity for SOFC electrolytes. In the last decade, thin film technology has helped to reduce the electrolyte thickness by more than oder of magnitude. With electrolyte thickness of around $10\,\mu m$, SOFCs can be operated efficiently at significantly lowered temperatures ($600...800\,°C$). In the case of a very thin electrolyte layer, either the anode or the cathode must serve as the structural

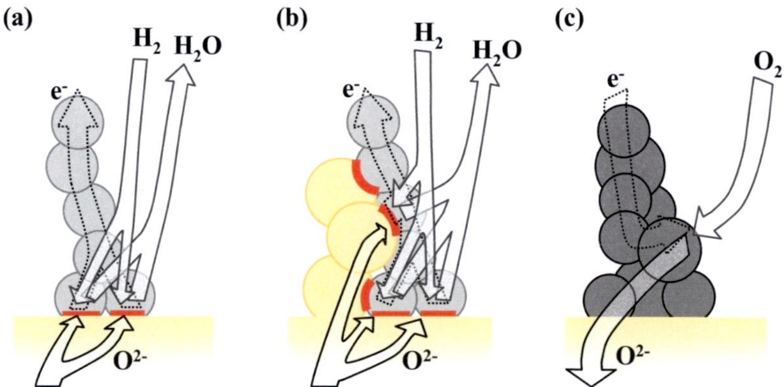

Figure 2.2.: SOFC electrode reactions and transport pathways for various electrode designs. (a) Metallic anode, (b) cermet anode and (c) MIEC cathode.

support for the MEA. Thereby, the so-called anode supported SOFC has proved to be successful. In the next paragraph, materials and design of anodes in anode supported cells (ASCs) are outlined briefly.

Anode

The principal function of SOFC anodes is to keep the anode reaction (Eq. 2.3) running at a high efficiency. This requires high catalytic activity and good transport properties for the transport of the involved reactants (i.e., O^{2-} ions, electrons, gaseous H_2 and H_2O). The latter implies high electronic and ionic conductivities combined with a high porosity. In ASCs, the anode structure must furthermore provide the mechanical strength necessary to give support to the MEA structure. Resulting SOFC anode design consists of various functional layers which differ in thickness, microstructure and chemical composition. The supporting element is a highly porous layer with thicknesses from 300 to 1500 μm.

Due to the reducing atmospheres ($pO_{2,an} \leq 10^{-18}$ atm), metals are preferably applied as electrode materials. The thermal expansion coefficient, which is required to match with that of the ceramic electrolyte, can be adjusted by mixing electrolyte material to the metal [29]. The resulting ceramic metal compound (cermet) offers electron conducting (metallic) phase and ion conducting (ceramic) phase. This has the advantage that the reaction zone of the electrode reaction (Eq. 2.3) is extended. The anode reaction involves O^{2-} ions, electrons and gaseous H_2 and H_2O. Figure 2.2a illustrates that the reaction hence can only proceed at the three phase boundary (TPB), where electron conducting phase, ion conducting phase and gas phase are available (marked in red). In cermet anodes, the TPB is extended spatially from the interface with the electrolyte towards the inner of the electrode structures, thus offering a higher density of electrochemically active area (cf. Fig. 2.2b). It should be noted that the design of the composite electrode structure has to guarantee for percolation of the three phases [30].

There are a number of reviews on anode materials available in literature [31–34]. The most commonly used and, hence, best-investigated material system is an anode cermet composed of metallic, electronic-conducting and catalytically active nickel (Ni) and ionic-conducting YSZ. Ni is comparatively cheap and Ni/YSZ anodes can be fabricated with conventional ceramic processing techniques (e.g., tape casting) at relatively low cost [15, 35]. Another advantage of Ni/YSZ anodes is the high catalytic activity of Ni with regards to the electrochemical fuel oxidation (Eq. 2.3). The high catalytic activity further enables conversion of hydrocarbon fuels at the SOFC anode (cf. Section 2.6). A disadvantage of Ni/YSZ anodes is the relatively fast kinetics of Ni oxidation at the usually high operating temperatures of the SOFC [36, 37]. Nickel tends to oxidize at high steam partial pressures, which are prevalent at the anode, when the cell is operated with a high fuel utilization. Apart from the deactivation as a catalyst, the transition from Ni to NiO implicates a substantial volume increase. The latter generates mechanical stresses, which involve the danger of a fatal damage of the cell assembly.

For the sake of completeness, it should be pointed out that recently, extensive research activities have focused on metal-oxide based anodes [38–40] as well as metal-supported anodes [41–46]. The first substitutes conventional cermet anodes by $SrTiO_3$, which offers the advantage of an enhanced re-oxidation reliability of the anode material. Adequate power densities are achieved when a catalytically active material, such as Ni and Pd, is added. Also metal-supported SOFC anodes offer potentially higher redox stability than conventional cermet anodes. Compared to cermets, metallic substrates are cheaper and more robust towards mechanical and thermal stresses. This makes metal-based SOFC stack technologies attractive for auxiliary power unit (APU) applications.

Cathode

The most common SOFC cathodes are ABO_3-type perovskite manganites, cobaltates, and ferrates (A = La, Sr, Ca; B = Mn, Co, Fe). Depending on the configuration, these perovskites can exhibit pure electronic but also ionic conductivity. Further important properties of the cathode material, such as chemical stability, thermal expansion coefficient and eventually electrochemical performance are basically depending on the composition of the perovskite. Detailed discussion of the resulting individual material properties for a broad variation of compositions can be found in literature [47–50].

Among those perovskites, lanthanum strontium cobalt ferrite (LSCF) represents the state of the art cathode material applied for a wide range of operating temperatures [47]. LSCF is a mixed ionic electronic conducting (MIEC) cathode material. This means that in principle, the entire surface of the LSCF cathode can act as TPB, as depicted schematically in Fig. 2.2c. It has to be noted that, for LSCF cathodes, direct contact with zirconia based electrolytes (such as YSZ) causes formation of an insulating $SrZrO_3$ interlayer [8, 51, 52]. Therefore, a gadolinium doped ceria (GDC) interlayer has to be applied between LSCF cathode and electrolyte [53].

In this thesis, cathodes made of LSCF with a stoichiometry of $La_{0.58}Sr_{0.4}Co_{0.2}Fe_{0.8}O_{3-\delta}$ are applied. This material system has been extensively studied [47, 53–56]. The initial performance of this cathode is very good, however it is important to know that the applied LSCF cathodes exhibit a pronounced and unsteady degradation. This behavior is well understood and detailed data for the degradation during the first $1000\,h$ of operation for various operating temperatures can be found in Refs. [50, 57, 58].

2.3. C/V Characteristics

The operation of real SOFC systems is characterized by various irreversibilities. In order to promote the development of the SOFC, a high interest lies on the understanding of these

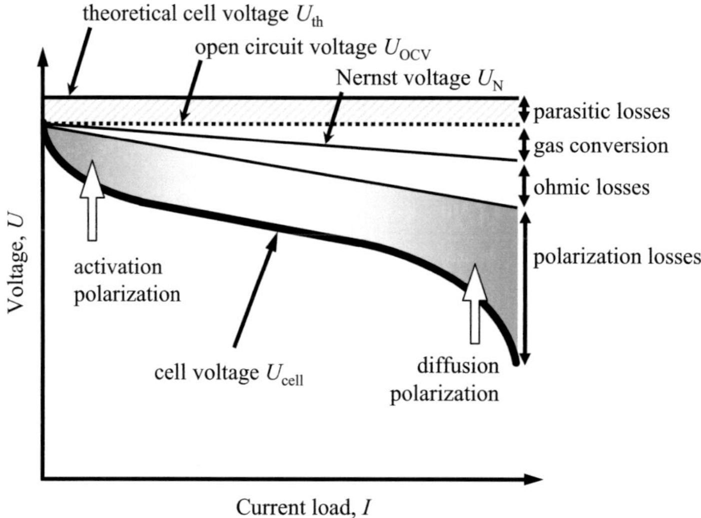

Figure 2.3.: Schematic representation of the C/V characteristics of a SOFC (adapted from Ref. [6]). With increasing current load, the individual loss processes cause a decreased cell voltage.

irreversibilities. Since the loss processes occurring during operation of an electrochemical device occur as overvoltages, electrochemical measuring techniques represent an obvious approach to the understanding of the irreversibilities. The most common electrochemical measuring technique applied in SOFC research is the current/voltage (C/V) measurement, which will be explained in the following.

In C/V measurements, a steady-state load current is applied to the cell and the resulting cell voltage is measured. The effect of the different loss mechanisms on the actual voltage output of a real SOFC for varying current densities is shown qualitatively in Fig. 2.3. The extent of the individual loss contributions is chosen freely [6]. As can be seen, even at open circuit condition (OCC), the measured cell voltage (U_{OCV}) is lower than the theoretical Nernst voltage U_{th} predicted by Eq. 2.4. This gap may be caused by various parasitic losses, like for example undesired electron leakage across the electrolyte or gas leakage of the system. The latter affects the species concentrations in fuel and oxidant, thus altering the chemical potential difference between each. A gap between theoretical and measured cell voltage at open circuit voltage (OCV) may further occur, when the fuel gas mixture is not in chemical equilibrium at the electrochemically active sites at the anode.

When a current load is applied to the cell, the measured cell voltage experiences a further decrease. Parts of this decrease have to be ascribed to the so-called gas conversion loss. Under operation of the fuel cell, the applied current density converts parts of the fuel and the oxidant. The fuel conversion at the anode implicates a decrease of the H_2 concentration along with an increase of the H_2O concentration. At the cathode, parts of the O_2 are consumed, resulting in a decreased pO_2. According to the modified Nernst equation (Eq. 2.7), U_N, decreases. It has to be pointed out that this voltage decrease is immanent in operation of the cell and has to be regarded as fuel utilization rather than as an actual loss process. In order to analyze the latter, the impact of the fuel utilization on the C/V curves can be minimized by raising the volumetric flow rates of fuel and oxidant gases.

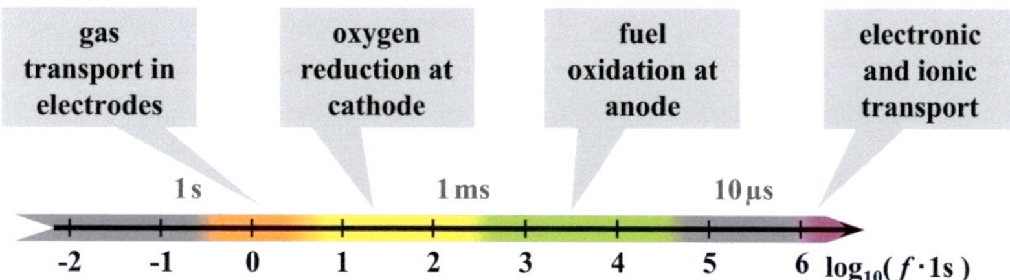

Figure 2.4.: Dynamics of a SOFC: Characteristic relaxation frequencies f of the individual electrochemical loss mechanisms occurring under operation of an anode supported SOFC.

The actual loss processes of the cell are considered to originate the characteristic S-type shape of the C/V curves measured on a SOFC. The losses are classified as (i) ohmic losses and (ii) polarization losses. The ohmic losses are caused by the non-ideal ionic and electronic transport through electrolyte and electrodes, respectively, as well as contact resistances between the single cell components.

The polarization losses can be further classified according to their physical origin as related to the charge transfer reactions at the electrodes (activation polarization) and to gas transport within the electrode structures (concentration polarization or diffusion polarization). In contrast to the ohmic losses, the resistance of these processes is dependent on the applied load current. Accordingly, the polarization losses show nonlinear C/V characteristics [7, 59, 60], and are hence considered as main representative of the characteristic shape of C/V curves measured on SOFCs.

The discussed loss processes exhibit different dependencies on the operating parameters (temperature, concentrations of H_2 and H_2O in the fuel and O_2 concentration in the oxidant gas) and thus a variation of the operation parameters results in different markedness of the characteristic shape of the measured C/V curves [7, 59, 60]. However it has to be noted that with the C/V measurement, only the sum of the prevailing loss contributions can be measured. For further separation of the individual loss contributions during operation of the SOFC, Electrochemical Impedance Spectroscopy (EIS) is a helpful tool, which will be introduced in the next section.

2.4. Electrochemical Impedance Spectroscopy

The significance of EIS relies on the fact that it reveals both the relaxation times[1] and the relaxation amplitudes of the various processes present in a dynamic system over a wide range of frequencies [61]. This makes this tool very attractive for the analysis of heterogeneous electrochemical material systems like the SOFC. Such systems are characterized by the simultaneous appearance of various electrochemical loss processes, which however are proceeding at different velocities. Figure 2.4 gives a schematic overview of the time constants of the electrochemical loss processes that determine the performance of the SOFC.

---

[1]The term relaxation generally refers to the return of a system into an equilibrium state as a response to a perturbation signal. The time constant of the relaxation of a physical process is called relaxation time, τ_0. Accordingly, the relaxation frequency is $f_0 = 1/(2\pi\tau_0)$.

An essential distinguishing characteristic of the electrochemical loss processes of the SOFC is that they exhibit different characteristic time constants, which are spread over several orders of magnitude (cf. Fig. 2.4).

- **Gas transport in the electrodes:** The mainly diffusive transport of the gas species in the porous electrode structures is the slowest loss mechanism. Depending on the gas composition, thickness and microstructure of the electrodes, the characteristic relaxation frequencies are in the range of $0.3\dots10$ Hz.

- **Cathode reaction:** The catalytically activated oxygen reduction at the cathode (Eq. 2.2) is faster than gas transport. Mainly depending on the catalytic activity of the cathode material and on the operating temperatures, the characteristic relaxation frequencies of the cathode reaction lie in the range of $2\dots500$ Hz.

- **Anode reaction:** Due to the high catalytic activity of the Ni, the processes related to the electrochemical fuel oxidation at the anode (Eq. 2.3) proceed faster than the cathode reaction. Depending on the catalytic activity of the anode and the operating temperatures, the characteristic relaxation frequencies lie generally in a range of $0.2\dots50$ kHz.

- **Ion and electron conduction:** Charge transport in electrolyte and electron conducting phases of the MEA are much faster than all the aforementioned processes. At the typical operating temperatures of the SOFC (above $600\,^{\circ}$C), the corresponding relaxation frequencies are beyond 1 MHz.

In contrast to steady-state C/V measurements, the transient EIS makes use of the differing relaxation times of the individual loss processes. Generally, the system is excited by a periodic small-signal perturbation at defined stimulus frequencies. In theory, sweeping of the stimulus frequencies results in a separated excitation of the individual electrochemical loss processes due to their different characteristic relaxation frequencies. The measurable response of the system hence reveals its dynamics, providing a higher resolved information about the excited electrochemical loss processes.

In recent years, EIS has been established as one of the most important non destructive characterization methods for electrochemical systems such as batteries or fuel cells. Combined with adequate analysis tools, it has the potential to provide overall understanding of the physical loss processes that form the internal resistance of the analyzed electrochemical device. Several comprehensive textbooks about impedance analysis and its application in electrochemistry are available in literature [62–64]. An introduction to the application of EIS in SOFC research can be found in Ref. [61].

2.4.1. Technical Realization

The most common and standard approach to measure the dynamic behavior (impedance) of an electrochemical system is by perturbating the system and measuring phase shift and amplitude of the system response via the measuring variables voltage and current. According to the way, the sinusoidal perturbation signal is provided, the measuring techniques are categorized as follows.

- **Galvanostatic:** In this case, the perturbation is realized by applying an alternating current with defined amplitude and frequency. This method is especially appropriate for small impedance values. When the impedance further decreases during the measurement, the peak current is limited by definition.

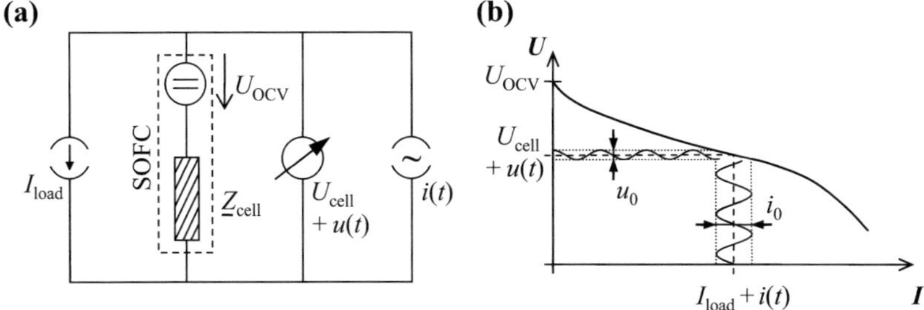

Figure 2.5.: (a) Electrical circuit of the measurement setup for the impedance measurement on a SOFC single cell with the internal impedance $\underline{Z}_{cell}$. (b) Illustration of the impedance measurement with the corresponding C/V curve. A sinusoidal current of small amplitude $i(t)$ is superposed to a defined bias current I_{load} and the voltage response $u(t)$ is measured [7, 65].

- **Potentiostatic:** Voltage perturbation is chosen for high impedances. The optimum perturbation amplitude – mostly $u_0 = 5 \dots 12\,\mathrm{mV}$ – is applied and the current response is measured. For high impedances, there is hence no danger of unintentionally high overvoltages at the system components.

The basic experimental arrangement for impedance measurements on a SOFC single cell is shown in Fig. 2.5a. A sinusoidal current of small amplitude $i(t) = i_0 \sin(\omega t)$ at a fixed angular frequency $\omega = 2\pi f$ is superposed to a defined bias current I_{load}. The resulting voltage response is measured, from which the sinusoidal alternating component $u(t) = u_0(\omega) \sin\left[\omega t + \varphi(\omega)\right]$ can be determined (Fig. 2.5b). From the ratio between the complex variables of voltage and current, the complex impedance $\underline{Z}$ for the frequency ω is calculated as follows [7, 65]:

$$\underline{Z}(\omega) = \frac{u_0(\omega)}{i_0} \cdot e^{j\varphi(\omega)} = |\underline{Z}(\omega)| \cdot e^{j\varphi(\omega)} = Z' + jZ''. \tag{2.8}$$

In Eq. 2.8, φ represents the frequency dependent phase shift between voltage and current. Z' and Z'' denote the real and the imaginary part of the complex impedance, respectively. It should be noted that an impedance is only defined for systems that satisfy the following conditions [63, 65, 66]:

i) **Causality**: A system is causal, when the measured response signal at any point of time exclusively depends on the perturbation signal at this point of time and/or its evolution until this point of time.

ii) **Linearity**: The measured response is a linear function of the perturbation signal, i.e., the relation between output and input underlies the principles of superposition and amplification.

iii) **Time-invariance**: The output of a time-invariant system does not depend explicitly on time, i.e., the system response on a certain perturbation signal should be exactly the same for any shift of time.

Although many electrochemical systems, including the SOFC, are usually non-linear, linearity can be assumed when the magnitude of the applied current stimulus is small enough

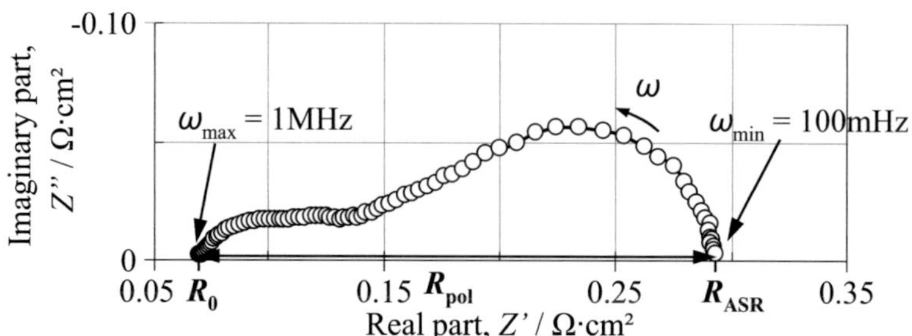

Figure 2.6.: Typical Nyquist-plot of a complex electrochemical impedance spectrum recorded on an anode supported SOFC single cell.

to cause a linear voltage response [7,67]. The required time-invariance implicates that the applied measuring samples should not change their electrochemical properties during the measurement. In this work, the state of the art ASCs manufactured by Forschungszentrum Jülich (FZJ) were applied as testing samples (cell details in Section 3.1). With reported degradation rates below $10\ m\Omega cm^2/kh$ [68], the FZJ ASC is one of the most stable and most reproducible cell systems available. Time-invariance of the samples during impedance measurements, which take about 30 min is hence guaranteed.

An impedance measurement is generally performed for a discrete quantity of frequency values ω_n in a frequency range $\omega_{min} \leq \omega_n \leq \omega_{max}$ relevant to the analyzed system. Expediently, the frequency range is swept in a logarithmic mode, i.e., the measuring frequencies are chosen in a way that every decade has the same amount of measuring frequencies. The recorded impedance values are usually plotted in the complex plane. This representation is also known as Nyquist-plot.

Figure 2.6 gives an example for a Nyquist-plot of an electrochemical impedance spectrum measured on an anode supported SOFC single cell. At very low frequencies (ω_{min}), all electrochemical loss processes are excited, thus contributing to the measured impedance. Since the perturbation frequency is much lower than the relaxation frequencies of the individual loss processes, the voltage response is in phase with the current perturbation and the imaginary part of the recorded impedance is zero. The intercept at low frequencies is hence identical with the differential resistance, which can be obtained from the corresponding C/V curve at the given operating point.

Sweeping to higher perturbation frequencies, the perturbation signal becomes faster than some of the excited loss processes. The slower loss processes begin to lag behind the perturbation signal and they decrease in the amplitude of their perturbation response. This results in a phase shift between excitation current and measured voltage response and in a decrease of the response amplitude. In this frequency region, the imaginary part of the impedance is hence unequal zero, while the real part decreases. Once the excitation signal is much faster than the relaxation frequency of a physical process, this process is not excited anymore. Hence, with increasing ω, an increasing number of physical loss processes do not further contribute to the measured impedance. At very high frequencies, no physical loss processes are excited anymore, resulting in a voltage response, which is again in phase with the current perturbation. The high-frequency intercept with the real axis corresponds to the ohmic resistance R_0 of the cell. The difference between the low and high-frequency intercept is the so-called polarization resistance R_{pol} of the cell. R_{pol}

is the sum of all single polarization resistances caused by the individual loss mechanisms.

EIS measurements applied on the SOFCs enables a clear separation of the ohmic resistance and the polarization resistance of the cell. However, a separated analysis of the physical processes themselves cannot be performed directly from the measurement data if their impedance contributions overlap in the spectrum. Therefore, the impedance data has to be analyzed with respect to the underlying dynamic processes. In the following sections, the impedance analysis methods applied in this work are presented.

2.4.2. Impedance Analysis

Equivalent Circuit Fitting

An equivalent circuit model (ECM) tries to reproduce the impedance response of an electrochemical system based on a composition of electrical equivalents. In principle, it is possible to accurately describe the impedance response of an electrochemical system with a sufficiently high number of passive elements, for as the system can be considered as causal, linear and time invariant during the measurement (cf. Section 2.4.1).

Impedance analysis by means of a physically motivated ECM has the potential to quantitatively analyze the individual loss processes, that determine the resistance of the investigated electrochemical system. Generally, this requires knowledge of (i) the amount of the contributing physical loss processes and (ii) their physical origin or, at least, the assignment to a dominating physical mechanism. In both questions, the distribution of relaxation times (DRT) method is of great help, which will be discussed in the following paragraph. Further, the individual elements in ECM are required to accurately describe the impedance of the corresponding loss process with respect to its physical origin. A detailed description of the most common equivalent circuit elements to model electrochemical systems is given in Refs. [7, 69].

With a physically adequate ECM set up, the impedance of an electrochemical system can be accurately described as the sum of the impedances of the individual equivalent circuit elements representing the physical loss processes of the system. The single equivalent circuit elements are characterized by parameters that bear substantial information about the represented loss process, such as

- its area specific resistance (ASR),

- its characteristic time constant,

- the degree of its time-constant dispersion[2].

The parameterization of an ECM is commonly done by fitting the parameters of a pre-designed equivalent circuit to impedance data obtained from measurement. Most commonly, this is performed by means of a complex nonlinear least squares (CNLS) fitting algorithm. The aim of the least squares fitting procedure is to find a set of parameters which will minimize the sum of the quality criterion as given by [62]:

$$S = \sum_{n=1}^{N} \left[w_{\mathrm{Re},n} \left(Z'(\omega_n) - Z'_{\mathrm{mod}}(\omega_n, \mathbf{a}) \right)^2 + w_{\mathrm{Im},n} \left(Z''(\omega_n) - Z''_{\mathrm{mod}}(\omega_n, \mathbf{a}) \right)^2 \right], \quad (2.9)$$

---

[2]By contrast to ideal processes with a defined characteristic time constant, real processes exhibit a time-constant distribution due to the finite spatial extension of any real system with spatially distributed physical properties.

with the complex model fit-function as

$$\underline{Z}_{\mathrm{mod}}(\omega_n, \mathbf{a}) = Z'_{\mathrm{mod}}(\omega_n, \mathbf{a}) + jZ''_{\mathrm{mod}}(\omega_n, \mathbf{a}) \tag{2.10}$$

and the measured complex impedance as

$$\underline{Z}(\omega_n) = Z'(\omega_n) + jZ''(\omega_n). \tag{2.11}$$

In Eq. 2.9, the vector a contains all free parameters of the fit problem. The factors $w_{\mathrm{Re},n}$ and $w_{\mathrm{Im},n}$ are the weighting factors associated with the n-th data point.

In practice, the initial values set for the free fit parameters are an essential condition for the fit to converge in a global minimum. Also here, the DRT method is of great help in order to identify accurate initial values, as will be shown in the next paragraph.

Distribution of Relaxation Times

The distribution of relaxation times (DRT) is an electrochemical impedance analysis method developed at Institut für Werkstoffe der Elektrotechnik (IWE) [65]. Calculation of the DRT represents a demonstrated valuable approach to the challenge of finding an adequate ECM able to describe the physical behavior of SOFC single cells [7]. In the following, the concept of the DRT will be introduced.

The DRT method is based on the fact that every impedance spectrum, that was measured in compliance with the above mentioned criteria of causality, linearity and time invariance, can be represented by a sufficiently large number of resistor-capacitor (RC) elements in series [62]. For an infinite number of RC elements[3] with continuously increasing relaxation times τ from 0 to ∞, the impedance $\underline{Z}(\omega)$ can be expressed by the following integral equation containing the distribution function $\gamma(\tau)$:

$$\underline{Z}(\omega) = R_0 + R_{\mathrm{pol}} \int_0^\infty \frac{\gamma(\tau)}{1 + j\omega\tau} \mathrm{d}\tau, \tag{2.12}$$

with

$$\int_0^\infty \gamma(\tau)\mathrm{d}\tau = 1, \tag{2.13}$$

where $\omega = 2\pi f$ is the angular frequency and j is the imaginary unit. Equation 2.12 represents any measured impedance spectrum with an ohmic resistance R_0 and a total polarization resistance R_{pol}, such as the one shown in Fig. 2.6. It should be noted that this general model does not contain any physical meaning, it only reflects the system dynamics of the measured impedance spectrum. The expression $\gamma(\tau)/(1 + j\omega\tau)\mathrm{d}\tau$ specifies the fraction of the overall polarization with relaxation times between τ and $\tau + \mathrm{d}\tau$. In practice,

---

[3]A resistor-capacitor (RC) element is the parallel connection of a resistor and a capacitor. The impedance of an RC-element is given by $\underline{Z}_{\mathrm{RC}}(\omega) = \frac{R}{1+j\omega\tau}$, with the characteristic relaxation time $\tau = RC$. A detailed explanation of this fundamental impedance element, which approximately represents most technical processes can be found in Ref. [69].

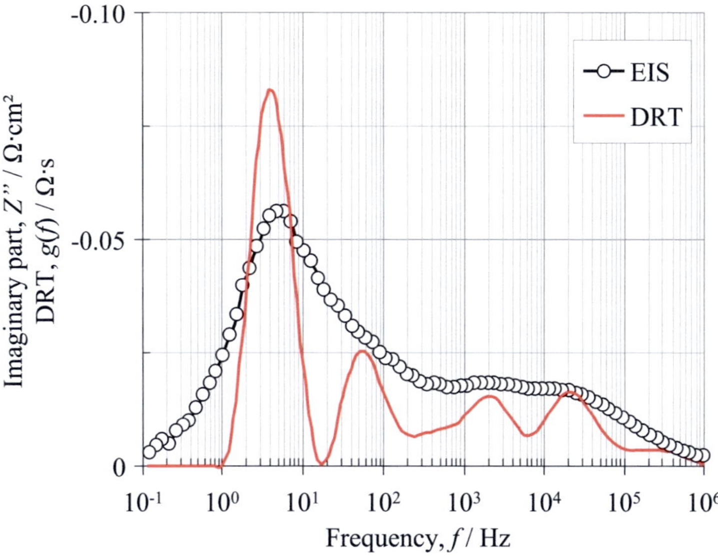

Figure 2.7.: Imaginary part of the measured impedance spectrum shown in Fig. 2.6 along with the corresponding Distribution of Relaxation Times (DRT, in red).

the continuous function $\gamma(\tau)$ is approximated by the discrete function γ_n for N serial RC elements with predefined, logarithmically distributed values for τ_n [69]:

$$\underline{Z}(\omega) = R_0 + R_{\text{pol}} \sum_{n=1}^{N} \frac{\gamma_n}{1 + j\omega\tau_n}. \tag{2.14}$$

Here, γ_n weights the contribution of the n-th RC element (with the relaxation time τ_n) to the overall polarization resistance. The calculation of γ_n represents an inverse ill-posed problem [70, 71]. The solution of inverse problems can be approximated numerically with the help of regularization. At the IWE, the Tikhonov regularization method [72, 73] employing the software package FTIKREG [74] is applied in order to calculate the distribution function from measured impedance data. This algorithm has been implemented in a user-friendly Microsoft Excel sheet with VBA macros by Dipl.-Ing. Volker Sonn (IWE). More detailed information about the DRT can be found in Refs. [7, 65, 75].

The benefit of this function is demonstrated in Fig. 2.7, where the negative imaginary part of the measured impedance spectrum given in Fig. 2.6 is compared to the distribution function calculated from the measured spectrum.[4] Unlike the imaginary part of the complex impedance curve, where the individual polarization processes overlap, four peaks can be distinguished clearly in the DRT. Each peak represents a physical polarization process. The characteristic relaxation frequency of each polarization process can be precisely determined by the corresponding peak frequency f_c in the DRT. The area enclosed by each

---

[4]In literature, it is distinguished between the analytically calculated distribution function $\gamma(\tau)$ and the numerically approximated $g(\tau)$ [65, 75]. In the following of this thesis, the numerically approximated DRT will be used to analyze measured EIS spectra. Since the DRT-function will be plotted versus relaxation frequencies f, the denomination $g(f)$ will be used.

peak corresponds to the ASR of the underlying polarization process. Furthermore, peak height $g(f_c)$ and the half-width h_g relative to the peak height give qualitative information about the time-constant dispersion of the underlying physical process [65].

The high resolution of physical processes in the DRT enables the exact identification of the individual polarization processes that form the impedance spectra of electrochemical devices. With this knowledge, physically meaningful ECMs can be set up. Furthermore, impedance analyses by means of the DRT provide (i) a precise determination of the characteristic frequencies of the individual polarization processes as well as (ii) qualitative information about the extent of their contribution to the overall polarization resistance. This delivers accurate initial values for the CNLS-fit (cf. Section 2.4.2). These characteristics are beneficial for the analysis of impedance spectra, which will be demonstrated in Chapter 4.

2.5. Electrochemical Loss Processes

In this section, an overview of the electrochemical loss processes that form the inner resistance of a SOFC is given. Since the cell type examined in this work is the state of the art ASC manufactured by FZJ (cf. cell details in Section 3.1), the following explanations are reduced to this cell type exclusively. For operation with H_2 as anode fuel, extensive knowledge about the electrochemical loss processes of this cell type has been achieved, which will be discussed in the following.

2.5.1. Ion Transport

In ideally contacted cells, ohmic losses occur during the electronic or ionic transport through the electrodes and the electrolyte. The overall ohmic resistance R_{ohm} is the sum of each individual ohmic contribution R_n. According to Ohm's law, the ohmic overpotential linearly increases with the current density I:

$$\eta_{\mathrm{ohm}} = I \cdot \sum_{n=1}^{N} R_n = I \cdot R_{\mathrm{ohm}}, \tag{2.15}$$

For the cells applied herein, the ionic conductivity of the 8YSZ electrolyte and the additional GDC interlayer is by orders of magnitude lower than the electronic conductivity of the electrodes [67]. Therefore, the ohmic resistance measured on ASCs has to be mainly ascribed to ion conduction through electrolyte and GDC interlayer. Oxygen ion conduction in the ceramic electrolytes is a thermally activated charge transfer mechanism. The characteristic temperature dependence of the oxygen ion conductivity is expressed by a modified Arrhenius equation

$$\sigma_{\mathrm{ion}} = \sigma_{0,\mathrm{ion}} \cdot \frac{1}{T} \cdot exp\left(\frac{-E_{\mathrm{act,ion}}}{RT}\right), \tag{2.16}$$

where σ_0 represents a material specific constant, $E_{\mathrm{act,ion}}$ is the activation energy, T is the absolute temperature and R is the gas constant. For YSZ, $\sigma_{0,\mathrm{ion}} \approx 3.6 \times 10^5$ S K/cm and $E_{\mathrm{act,ion}} \approx 8 \times 10^4$ J/mol [15].

2.5.2. Electrode Reactions

Activation polarization describes the electrochemical loss mechanisms taking place mainly at the TPB, where ionic-conducting, electronic-conducting and gas phase meet. Primarily, activation polarization refers to the irreversibilities of the charge transfer reactions taking place at the SOFC electrodes. These are mainly determined by the velocity of the reaction, which is dependent on the catalytic activity of the electrodes with respect to the corresponding charge transfer reaction.

As discussed in Section 2.2, design of technical electrodes (MIEC cathodes and composite anodes) offers a great number of electrochemically active sites resulting in a significant expansion of the electrochemically active zone into the volume of the electrode (cf. Fig. 2.2). Within this reaction zone, electronic, ionic, and gas transport occur in parallel pathways, delivering the charge transfer reaction with reactants. Since these processes are not ideal, a certain influence of these transport processes within the dimensions of the electrochemical reaction zone on the activation overpotential $\eta_{\mathrm{act,el}}$ has to be considered.

Steady-State Behavior

For the SOFCs applied in this work, the activation polarization overpotential $\eta_{\mathrm{act,el}}$ exhibits non-linear C/V characteristics over a broad range of operating parameters. It has been shown, that the Butler-Volmer ansatz is able to accurately describe this non-linear behavior [7,59,60]. The well-known Butler-Volmer equation implicitly describes the dependence of the anode activation potential $\eta_{\mathrm{act,el}}$ on the applied current density i as [63,76,77]

$$i = i_{0,\mathrm{el}} \left[\exp\left(\alpha_{\mathrm{el}} \frac{zF}{RT} \eta_{\mathrm{act,el}} \right) - \exp\left(-(1 - \alpha_{\mathrm{el}}) \frac{zF}{RT} \eta_{\mathrm{act,el}} \right) \right], \qquad (2.17)$$

where $i_{0,\mathrm{el}}$ denotes the exchange current density, z the number of transferred electrons (in the case of the SOFC: $z = 2$), α_{el} the apparent charge transfer coefficient, and $\eta_{\mathrm{act,el}}$ the activation overpotential of the corresponding electrode (anode or cathode). The charge transfer coefficient is an indicator of the symmetry of the activation energy barrier when a positive or negative overpotential is applied [76–78].

The exchange current density $i_{0,\mathrm{el}}$ is defined as the microscopic flux crossing the electrode/electrolyte interface equally in both directions at equilibrium. It reflects the kinetics of the catalytically activated electrode reaction. For SOFC electrodes, the following semi-empirical relations can be applied in order to describe the electrode kinetics [7,79]:

$$i_{0,\mathrm{cat}} = \gamma_{\mathrm{cat}} \cdot (pO_{2,\mathrm{cat}})^c \cdot \exp\left(\frac{-E_{\mathrm{act,cat}}}{RT} \right). \qquad (2.18)$$

$$i_{0,\mathrm{an}} = \gamma_{\mathrm{an}} \cdot (pH_{2,\mathrm{an}})^a \cdot (pH_2O_{\mathrm{an}})^b \cdot \exp\left(\frac{-E_{\mathrm{act,an}}}{RT} \right). \qquad (2.19)$$

The equations represent the thermal activation (Arrhenius behavior) of the electrode charge transfer reactions, characterized by the activation energy $E_{\mathrm{act,el}}$, and the characteristic dependencies on the partial pressures of the gaseous reactants($pO_{2,\mathrm{cat}}$, $pH_{2,\mathrm{an}}$ and pH_2O_{an}).

Transient Behavior

The design properties of technical SOFC electrodes resulting in distributed charge transfer reactions and transport processes (as explained in Section 2.2) come into effect, when the transient impedance response of the activation polarization is analyzed. Therefore, distributed circuit elements have to be considered for a detailed physical interpretation of the activation polarization impedance data taken on technical SOFC electrodes. The so-called transmission line models (TLMs) represent the spatially distributed coupling of charge transfer reactions with several reactant transport pathways [80]. Transmission lines for porous electrodes consist of several branches representing the individual transport pathways in the electrode. These branches, which are represented by a serial arrangement of infinite elements in a finite space, are connected mutually by impedance elements representing the charge transfer reaction. Detailed descriptions of TLMs can be found in literature [81, 82].

LSCF-Cathode For the MIEC (LSCF) cathode, it has been shown that the Gerischer element precisely describes the activation polarization impedance response [83]. Assuming negligible losses in electronic conduction, the Gerischer element is a TLM with two branches representing ionic and gas transport in the cathode. A detailed explanation of this model can be found in Ref. [84]. The analytical expression of the complex Gerischer impedance is given by

$$\underline{Z}_{\mathrm{G}}(\omega) = \frac{R_{\mathrm{chem}}}{\sqrt{1 + j\omega\tau_{\mathrm{chem}}}}, \tag{2.20}$$

where the characteristic resistance R_{chem} and the characteristic time constant τ_{chem} are related to the thermodynamic, surface kinetic, and transport properties of the mixed conductor. A series of publications demonstrates the successful application of this model for the accurate quantitative analysis of LSCF cathodes via EIS measurements on anode supported SOFCs. With the assistance of the Gerischer impedance model,

- the contribution of the degradation of LSCF cathodes to the overall cell degradation experienced by SOFCs [57],

- the surface exchange kinetics and oxygen ion bulk diffusion properties (k^{δ} and D^{δ}) of the MIEC material [58, 85] and

- the kinetic parameters (c, $E_{\mathrm{act,cat}}$ and γ_{cat}) and the cathodic charge transfer coefficient (α_{cat}) for the prediction of the cathodic exchange current density (Eq. 2.18) and the Butler-Volmer equation (Eq. 2.17) [59]

were determined accurately.

Ni/8YSZ-Anode The Ni/8YSZ cermet anode is a porous two-phase composite electrode. The two interpenetrating matrices, the ionic-phase (8YSZ) and the electronic phase (Ni), result in a great number of electrochemically active sites. Created by the high ionic conductivity of 8YSZ the electrochemically active zone significantly expands into the volume of the electrode.

Also in the research of Ni/YSZ SOFC anodes, transient analyses have contributed to deeper insights into the coupling of transport resistances and the distributed charge transfer reaction. For the anodes applied in this study, it is demonstrated that the impedance of the activation polarization process exhibits two clearly distinctive time constants [86, 87].

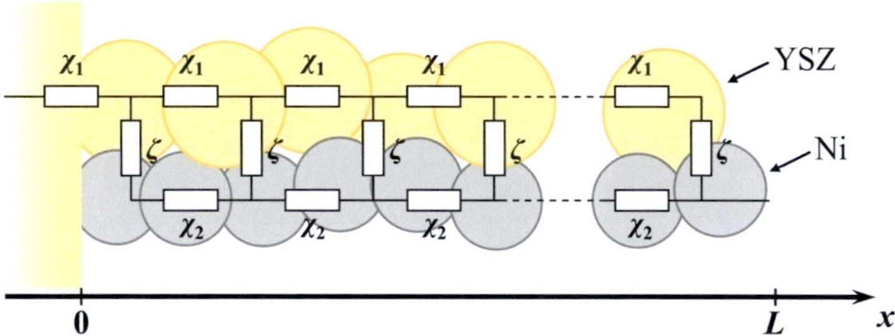

Figure 2.8.: Modeling scheme for a porous two-phase composite electrode. The transmission line model proposed by Sonn et al. accounts for the charge transport in ceramic phase (χ_1) and metallic phase (χ_2), as well as the charge transfer reaction (ζ) (adapted from Ref. [87]).

Sonn et al. have investigated the physical origin of these two semicircles with a combined DRT and ECM fitting analysis [87]. The impedance response of the anode activation polarization was modeled by a TLM, as depicted schematically in Fig. 2.8.

The overall impedance for the TLM with the geometric length L is expressed by the subelements χ_1, χ_2 (both ohmic resistances) representing the distributed resistance of the ionic conducting (YSZ) and electronic conducting (Ni) phase, respectively. The subelements ζ (RC elements), that connect both aforementioned pathways, are representing the impedance of the charge transfer reaction. With this combined DRT and ECM fitting analysis, the two characteristic time constants have been assigned to (i) the anodic charge transfer reaction at $2000\ldots8000$ Hz and (ii) ionic transport in the YSZ matrix at $20\ldots30$ kHz [87].

In the ECM for ASCs proposed by Leonide et al. [86], the activation polarization impedance is represented by two RQ elements[5]. It is shown that the impedance response of the anode activation polarization can be described accurately by the two RQ elements, which is not physically meaningful but more practical and sufficient for the analysis of the loss contribution of this process ($R_{\mathrm{act,an}}$) to the overall cell resistance.

With this method of determining $R_{\mathrm{act,an}}$, the penetration depth of the charge transfer reaction within the electrode structures has been analyzed for ASCs with varying thicknes of the anode functional layer (AFL) [88]. It has further been employed to analyze the kinetics of the anodic charge transfer reaction, as represented by the exchange current density in Eq. 2.19. The analysis of the obtained $R_{\mathrm{act,an}}$ for systematic variation of fuel gas composition and operating temperature has lead to the determination of the kinetic parameters a, b and $E_{\mathrm{act,an}}$ [59].

2.5.3. Gas Transport

Under operation of the cell, i.e., when a current load is applied to the cell, the reactants of the electrode reactions are transported through the electrode structures: In the cathode, O_2 is transported to the electrochemically active sites, where it is consumed. At the anode, H_2 is transported to the electrochemically active site, where it is converted into H_2O, which

---

[5]An RQ element is the parallel connection of a resistor and a constant phase element (CPE). Details can be found in [7, 62]

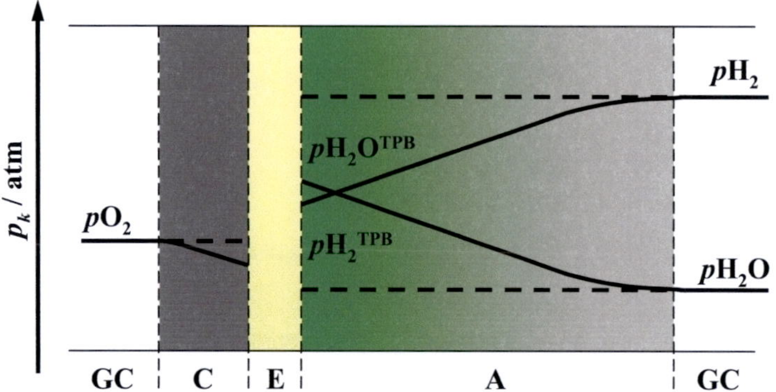

Figure 2.9.: Schematic representation of the partial pressure profiles within the electrode structure of an ASC (solid lines) that arise during operation of the cell (dashed lines represent partial pressures at OCV; GC = gas channel, C = cathode, E = electrolyte and A = anode).

in return has to be transported toward the gas channel. The predominant mechanism of these non-ideal transport processes is gas diffusion.

Steady-State Behavior

Generally, a diffusive species flux is inherently linked to a concentration gradient, as described in Fick's first law for the assumption of steady state [89]:

$$J_{d,k} = -D_k \nabla c_k. \tag{2.21}$$

According to Fick's law, the diffusion coefficient D_k describes the relation between diffusive flux $J_{d,k}$ and concentration gradient ∇c_k of species k. In the case of the SOFC, the diffusive species fluxes in the electrodes are corresponding to the amount of species that are converted in the electrode reactions. For operation with H_2 as fuel and O_2 as oxidant, the relation between the diffusive fluxes $J_{d,k}$ of the gas species and the applied current density i can hence be described by:

$$J_{d,H_2} = -J_{d,H_2O} = 2 \cdot J_{d,O_2} = \frac{RT}{2F} \cdot i. \tag{2.22}$$

The effect of gas transport and concentration profiles (represented in partial pressures $p_k = c_k \cdot p_0$, $p_0 = 1$ atm) in the case of the porous SOFC electrodes is depicted in Fig. 2.9. For sufficiently high amounts of fuel and oxidant gas flow rates, the species concentrations in the gas channels can be assumed as constant. Within the porous electrode structures, the applied current load and the diffusive species flux causes concentration gradients. This results in a depletion of O_2 at the electrochemically active sites in the cathode. At the electrochemically active sites of the anode, a depletion of H_2 along with an accumulation of H_2O arises.

For diffusion in the porous SOFC electrode structures, Fick's law (Eq. 2.21) can be specified as:

$$J_{d,k} = D_k^{\text{eff},el} \cdot \frac{c_k - c_k^{\text{TPB}}}{l_{el}}.$$
(2.23)

Equation 2.23 assumes linear concentration profiles extended over the entire electrode thickness l_{el}. The effective diffusion coefficient for the porous electrode structure can be described by

$$D_k^{\text{eff},el} = \psi_{el} \cdot D_{\text{mol},k},$$
(2.24)

where the parameter $\psi_{el} = \varepsilon_{el}/\tau_{el}$ represents the electrode microstructure and describes the ratio of porosity ε and tortuosity τ [90]. In SOFC electrodes, the molar diffusion coefficient $D_{\text{mol},k}$ must account for gas phase bulk diffusion as well as Knudsen diffusion [90, 91].

According to Eqs. 2.22 and 2.23, the extent of the resulting concentration gradients under operation of the SOFC is depending on the (i) value of the applied current density i, (ii) microstructural parameters of the electrode ψ_{el} and the molar diffusivity of the species inside the electrode microstructures and (iii) the molar species concentrations in the gas channel c_k.

The depletion and accumulation of species at the electrochemically active sites cause a deviation of the electrode potentials. The resulting overpotentials can be calculated with the Nernst equation (Eq. 2.7) [7, 76, 91, 92]. At the cathode, the resulting concentration overpotential η_{conc} is

$$\eta_{\text{conc,cat}} = \frac{RT}{4F} \cdot \ln \left(\frac{pO_{2,\text{cat}}}{pO_{2,\text{cat}}^{\text{TPB}}} \right)$$
(2.25)

and at the anode

$$\eta_{\text{conc,an}} = \frac{RT}{2F} \cdot \ln \left(\frac{pH_{2,\text{an}} \cdot pH_2O_{\text{an}}^{\text{TPB}}}{pH_2O_{\text{an}} \cdot pH_{2,\text{an}}^{\text{TPB}}} \right).$$
(2.26)

The resulting concentration overpotential corresponds to the Nernst potential that arises from the concentration gradients through the electrode thickness. Combining Eqs. 2.22, 2.23 and 2.25, 2.26, respectively, results in the following relations between applied current density i and electrode concentration overpotentials $\eta_{\text{conc},el}$ [7, 91]:

$$\eta_{\text{conc,cat}} = \frac{RT}{4F} \cdot \ln \left(\frac{1}{1 - \dfrac{RT}{4F} \dfrac{l_{\text{cat}} \left(1 - pO_{2,\text{cat}}/p_0\right)}{D_{O_2}^{\text{eff,cat}} \cdot pO_{2,\text{cat}} \cdot P} \cdot i} \right),$$
(2.27)

$$\eta_{\text{conc,an}} = \frac{RT}{2F} \cdot \ln \left(\frac{1 + \dfrac{RT}{2F} \dfrac{l_{\text{an}}}{D_{H_2O}^{\text{eff,an}} \cdot pH_2O_{\text{an}} \cdot P} \cdot i}{1 - \dfrac{RT}{2F} \dfrac{l_{\text{an}}}{D_{H_2}^{\text{eff,an}} \cdot pH_{2,\text{an}} \cdot P} \cdot i} \right).$$
(2.28)

The parameter p_0 denotes the absolute pressure and $P = 101330\,\text{Pa/atm}$ is a conversion factor from [atm] to [Pa]. Equations 2.27 and 2.28 describe the C/V characteristics of the electrode concentration polarization. As can be seen, the C/V characteristics depend on (i) the microstructural parameters of the electrode ψ_{el} and the molar diffusivity of the species inside the electrode microstructures, as well as (ii) the species concentrations at the gas channel.

In the following paragraph, it will also be shown how the transient analysis of the diffusion polarization can be used in order to determine microstructural parameters of the SOFC anode.

Transient Behavior

In this paragraph, only the transient behavior of the diffusion polarization in the SOFC anode will be analyzed. In ASCs operated with air as oxidant, it has been shown that the cathode diffusion polarization plays a negligible role in the overall polarization losses that form the measured impedance spectra at OCV [86].[6]

The complex impedance of the gas diffusion polarization in SOFC anodes is based on the transient concentration profiles that evolve during transient perturbation of the system. Transient gas diffusion is described by Fick's second law [89] as

$$\frac{\partial c_k}{\partial t} = \nabla \cdot \left(D_k \nabla c_k \right), \tag{2.29}$$

where c_k denotes the concentration of species k and t the time. Assuming (i) high volumetric gas flow rates of oxidant and fuel in the gas channel and (ii) transient perturbation at OCV, the diffusion becomes one-dimensional with x indicating the direction of the thickness of the porous electrode structure. With the molar diffusion in SOFC anodes as described above, Eq. 2.29 can hence be specified as:

$$\frac{\partial c_k}{\partial t} = D_k^{\text{eff},el} \frac{\partial^2 c_k}{\partial x^2}. \tag{2.30}$$

It is further assumed that the charge transfer reaction in the electrode (activation polarization processes, cf. Section 2.5.2) proceeds infinitely fast. This implicates that exclusively the species concentrations in the anode are excited at the interface electrode/electrolyte ($x = 0$) by the applied stimulus current density i:

$$\frac{\partial c_k}{\partial x} = \frac{1}{2FD_k^{\text{eff},el}} \cdot i \quad \text{for} \quad t \geq 0. \tag{2.31}$$

At the interface electrode/gas channel ($x = l_{el}$), fixed species concentrations are given as boundary condition:

$$c_k = c_k(0) \quad \text{for} \quad t \geq 0. \tag{2.32}$$

From these equations, the so-called general finite-length Warburg (GFLW) impedance can be deduced [62, 93]:

---

[6]This is due to (i) the comparably small thickness of the cathode ($l_{\text{cat}}/l_{\text{an}} < 0.05$ for the cells applied in this work, cf. Section 3.1) and (ii) the relatively flat C/V characteristics (Eq. 2.27) at $pO_{2,\text{cat}} = 0.21$ atm.

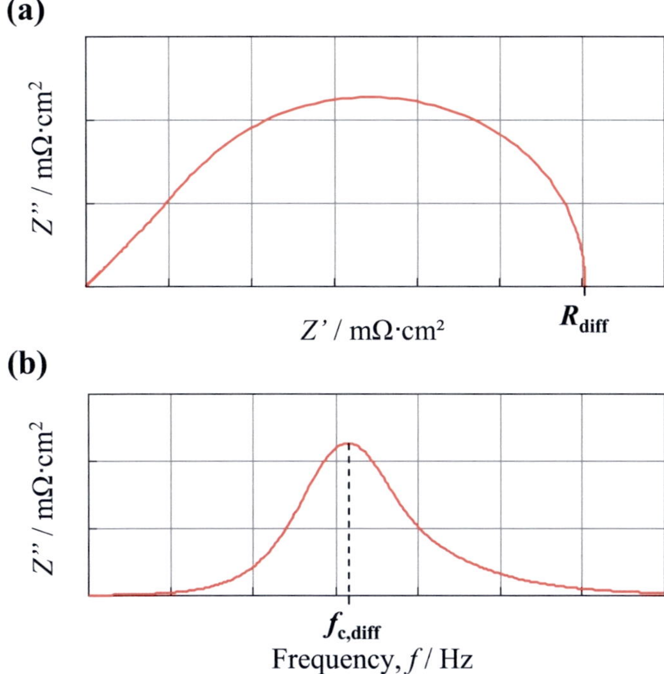

Figure 2.10.: Concentration polarization impedance due to a finite-length diffusion process (Warburg diffusion) as typical impedance response of the fuel gas transport in anode supported SOFCs. (a) Nyquist representation and (b) imaginary part vs. frequency.

$$\underline{Z}(\omega)_{\text{diff},el} = R_{\text{diff},el} \frac{\tanh l_{el} \sqrt{\left(j\omega / D_k^{\text{eff},el} \right)}}{\sqrt{\left(j\omega / D_k^{\text{eff},el} \right)}}. \tag{2.33}$$

The characteristic frequency of the Warburg diffusion impedance can be approximated to [62, 91]

$$f_{\text{c,diff}} = \frac{2.53 \cdot D_k^{\text{eff},el}}{2\pi l_{el}^2}. \tag{2.34}$$

The finite-length Warburg impedance represents the impedance response of the anodic gas transport in anode supported SOFCs [86, 91]. The Nyquist representation of this impedance behavior is depicted in Fig. 2.10.

The ASR of the Warburg diffusion process in SOFC anodes is given by [59, 91]:

$$R_{\text{diff,an}} = \left(\frac{RT}{2F} \right)^2 \frac{l_{\text{an}}}{\psi_{\text{an}}} \left(\frac{1}{D_{\text{mol},H_2} \cdot pH_{2,\text{an}}} + \frac{1}{D_{\text{mol},H_2O} \cdot pH_2O_{\text{an}}} \right). \tag{2.35}$$

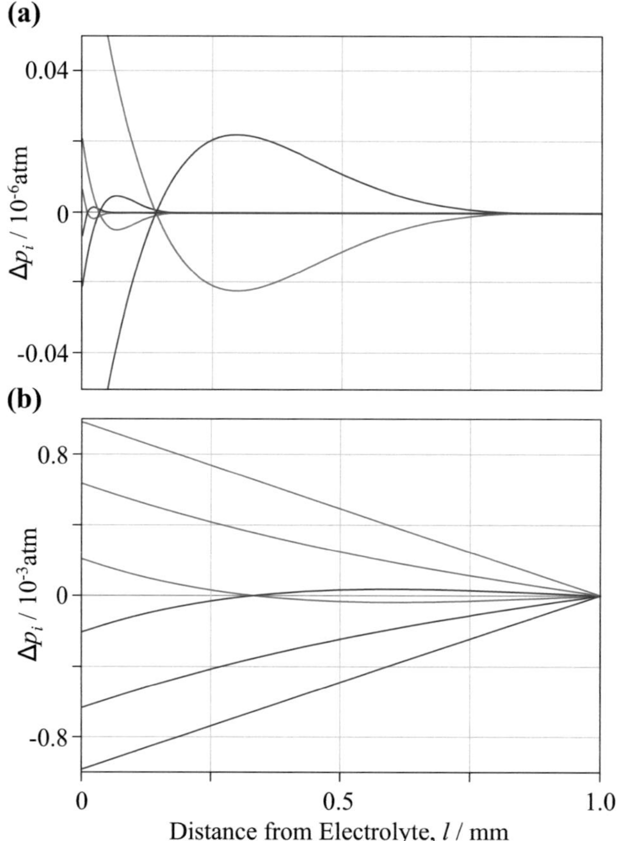

Figure 2.11.: Concentration profiles of H_2 (blue lines) and H_2O (green lines) through the thickness of the anode substrate for excitation at (a) high and (b) low frequencies. The depicted concentration profiles correspond to the maximum of the voltage response.

It has been demonstrated that, according to Eq. 2.35, the structural parameter ψ_{an} can be determined from EIS measurements under variation of $pH_{2,an}$ and pH_2O_{an} [59].

In order to understand the appearance of the Warburg impedance in the Nyquist plot, the underlying gas transport in SOFC anodes is analyzed. Figure 2.11 depicts the transiently excited concentration profiles in the electrode structure for various perturbation frequencies. The temporal evolution of the resulting diffusion overpotential $\eta_{conc,an}$ is depicted in Fig. 2.12.

The transient concentration profiles in the anode and the resulting overpotential responses show two different characteristic behaviors, depending on the frequency of the perturbation signal.

High frequencies: At very high frequencies, gas transport is rather slow and concentration gradients can only be observed at the electrochemically active sites. With decreasing stimulus frequencies ω, the concentration gradients at the interface anode/electrolyte

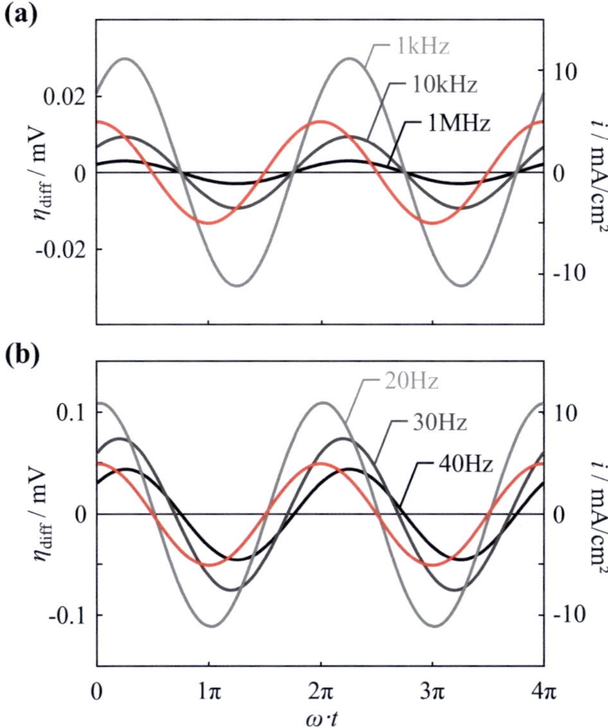

Figure 2.12.: Oscillating concentration overpotentials (gray scale lines) along with the stimulus current (red line) over dimensionless time for (a) high and (b) low frequencies.

arise. As depicted in Fig. 2.11a, concentration profiles that describe the form of an exponentially damped wave with a finite propagation length in the x-direction can be observed [76]. With decreasing stimulus frequency, the profiles show increasing penetration depths towards the gas channel. At high frequencies, the concentration perturbation however does not extend to the entire thickness of the anode structure. Therefore, the behavior of the concentration profiles is characteristic for diffusion in a stagnant medium of infinite dimension [63].

When looking at the resulting overpotentials at high frequencies (cf. Fig. 2.12a), it is observed that the amplitude increases with decreasing perturbation frequency. It is characteristic for high frequencies that the phase-shift between stimulus current and resulting overpotential is constant at $45°$. Accordingly, real and imaginary part of the complex impedance $\underline{Z}(\omega)$ have the same size. The parallel increase of Z' and Z'' is reflected in the characteristic $45°$ slope of the high-frequency part of the spectrum of $\underline{Z}(\omega)_{\text{diff}}$, which is depicted in Fig. 2.10.

Low frequencies: For lower frequencies, the concentration profiles extend to the thickness of the anode structure, where the concentration is kept constant (as specified in Eq. 2.32). Accordingly, the behavior at low frequencies is characteristic for gas transport in a finite-length diffusion layer. As depicted in Fig. 2.11b, the concentration gradients through the electrode thickness increase with decreasing stimulus frequencies toward a

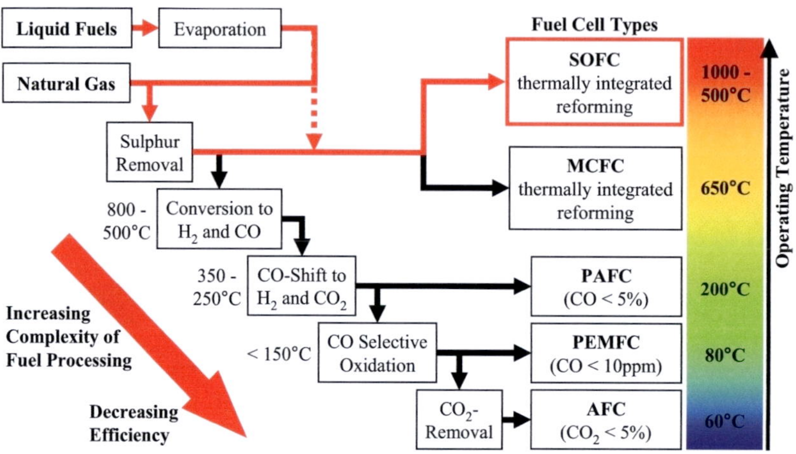

Figure 2.13.: Fuel processing steps (from hydrocarbon fuels to hydrogen) for different types of fuel cell systems (MCFC, molten carbonate fuel cell; PAFC, phosphoric acid fuel cell; PEMFC, proton exchange membrane fuel cell; AFC, alkaline fuel cell; adapted from Steele [94]).

linear concentration profile.

Accordingly, the resulting diffusion overpotential increases with decreasing frequencies (cf. Fig. 2.12b). However can be observed that the phase shift between stimulus signal and concentration overpotential decreases with decreasing frequencies. This means that the real part of the impedance increases to a higher extent than the imaginary part. Correspondingly, the EIS spectrum now follows the form of a semicircle (cf. Fig. 2.10). For the linear concentration profiles at very low frequencies, steady-state has been reached and the overpotential is in phase with the stimulus current. The imaginary part of the impedance hence tends toward zero.

2.6. Fuel Flexibility

The ideal fuel for any fuel cell system is considered to be hydrogen, which is planned to be generated from regenerative sources in the future. To the present day however, industrial hydrogen is almost exclusively generated from hydrocarbons. Due to the lack of a widespread hydrogen distribution infrastructure and to difficulties in its storage, it can be assumed that fuel cell systems have to rely on conventional fuels in the immediate future. Figure 2.13 gives a schematic overview of the various fuel processing steps, which have to be considered, when fuel cell systems are fueled with hydrocarbon-based fuels.

Regarding the efficiency of fuel cell systems under these premises, the fuel flexibility makes the SOFC the most attractive device. Due to the high operating temperatures, the high catalytic activity of the anode materials along with their low susceptibility, SOFC systems can directly use hydrocarbon or hydrocarbon-derived fuels. The range of possible fuels includes natural gas [94], bio-derived fuels [95], liquid fuels, such as propane, diesel, kerosene or alcohols [96–99], and coal gas [100]. At this point it should be noted that the operation on methane and natural gas is broadly applied, whereas no stable operation has been demonstrated for the direct use of any other hydrocarbon fuel, so far.

2.6.1. Hydrocarbon Conversion

The conversion of hydrocarbons into hydrogen or syngas[7] is a well-established industrial process. In SOFC systems, this catalytic fuel conversion can either proceed directly at the SOFC anode, which is called direct internal reforming (DIR) or in a thermally coupled upstream reformer via the so-called indirect internal reforming (IIR). Internal reforming in general is a broadly investigated topic in SOFC research and several reviews are available on this topic [15, 101–105]. The most relevant fuel processing steps for SOFC systems are described in the following.

Reforming

Steam reforming is the most common industrial process for producing hydrogen or syngas from hydrocarbon fuels or alcohols. Several comprehensive overviews can be found in literature [106–108]. For methane, the main reactions are the steam reforming reaction itself (Eq. 2.36) and the associated water-gas shift (WGS) reaction (Eq. 2.37). The so-called dry reforming (with carbon dioxide instead of steam, Eq. 2.38) acts in a similar way [109].

$$CH_4 + H_2O \rightleftharpoons CO + 3\,H_2, \qquad \Delta H^\circ = +206\,\text{kJ/mol} \tag{2.36}$$

$$CO + H_2O \rightleftharpoons CO_2 + H_2, \qquad \Delta H^\circ = -41\,\text{kJ/mol} \tag{2.37}$$

$$CH_4 + CO_2 \rightleftharpoons 2\,CO + 2\,H_2, \qquad \Delta H^\circ = +247\,\text{kJ/mol} \tag{2.38}$$

The high endothermic steam reforming reaction (Eq. 2.36) is thermodynamically favored at high temperatures and low pressures. In industry, the steam reforming reaction and the shift reaction (Eq. 2.37) are jointly carried out at temperatures typically above $700\,^\circ\text{C}$. Supported metal catalysts, usually nickel based, are applied.

Reactions 2.36 and 2.37 are reversible and at temperatures above $600\,^\circ\text{C}$, they proceed fast over active catalysts, including Ni/YSZ anodes. For SOFC systems with DIR, it can hence be assumed that the reactions 2.36 and 2.37 are in equilibrium for most parts of the system, with the product gases governed by the thermodynamic equilibria set up between both reactions.

Excess addition of steam moves the equilibrium toward more hydrogen, which however implicates a higher utilization of the fuel already in the processing step. Due to the requirement of steam, internal steam reforming is practical for SOFC systems in stationary applications.

Catalytic Partial Oxidation

The catalytic partial oxidation (CPO) of hydrocarbons is another well-established approach to generate syngas. Literature reviews on this process are given in Refs. [110,111]. In CPO the fuel is partially oxidized by a small amount of oxygen. For methane, the exothermic reaction is:

---

[7]Synthesis gas, or syngas, denotes gas mixtures of hydrogen and carbon monoxide.

$$CH_4 + \tfrac{1}{2}O_2 \rightleftharpoons CO + 2\,H_2, \qquad \Delta H^\circ = -247\,\mathrm{kJ/mol} \qquad (2.39)$$

For the CPO reaction, it is important that the oxygen is provided in an understoichiometrical amount, avoiding the combustion of the fuel. Nickel based catalysts are active and selective for the partial oxidation of methane.

The high utilization of the fuel is a disadvantage for the application in fuel cell systems, whereas the simplicity of the process, which does not require steam, is a key advantage. Therefore, this process is attractive in particular for APU applications, where it is implemented in terms of IIR.

Direct Electrochemical Conversion

There are certain differences concerning the definition of direct oxidation of hydrocarbons on SOFC anodes. A series of works report the operation of SOFC anodes on dry hydrocarbons without upstream reforming and without adding steam or air to the fuel. The results of such experiments show to what extent the anode can convert the dry hydrocarbon and the resistance or tolerance of the anode with respect to carbon formation. It is commonly reported throughout several operating temperatures ($700\ldots950\,^\circ\mathrm{C}$) that under application of moderate load currents, the Ni/YSZ SOFC anodes can be operated stable without coking, even with dry fuels [112–114]. The realization of stable operation with dry methane has been demonstrated in Ref. [112].

A further definition found in literature is the direct electrochemical oxidation of the fuels as [105]

$$CH_4 + O^{2-} \rightleftharpoons CO + H_2 + 2\,e^-, \qquad \Delta H^\circ \geq 0 \qquad (2.40)$$

A number of papers have been published claiming to report the direct electrochemical oxidation of methane and hydrocarbons in SOFCs anodes at relatively low temperatures (about $700\,^\circ\mathrm{C}$) [115, 116]. There is however strong evidence, that after an initial electrochemical oxidation, the formed species H_2 and CO are preferred in the electrochemical reaction and that the subsequent formation of H_2O or CO_2 promotes the reforming of the hydrocarbon species according to Eqs. 2.36 and 2.38, respectively [105].

Carbon Deposition

For the sake of completeness, it should be stated that the formation of carbon is a serious problem with many processes that involve hydrocarbons at high temperatures, including SOFCs. Since carbon formation is explicitly avoided in this work (cf. Section 3.3), this topic is not further addressed to herein. For general information on carbon deposition, the reader is referred to Refs. [107, 108]. Specific information on carbon deposition Ni/YSZ SOFC anodes can be found in the comprehensive works of Finnerty et al. [117] and Timmermann et al. [118].

2.6.2. Operation with Hydrocarbon Fuels

In SOFC literature, there is a great deal of knowledge on the catalytic heterogeneous conversion of hydrocarbon fuels achieved by both experimental and modeling analyses [15, 119–126]. However relatively little is known about the electrochemical processes, which determine the cell performance under operation with hydrocarbon fuels. An overall understanding of the electrochemistry of hydrocarbon fueled SOFC anodes comprises detailed knowledge of the following aspects:

- The mechanism of the anodic charge transfer reaction,

- the transport pathways for the reactants involved in the anode reaction and

- the impact of reforming chemistry on the former two aspects.

Thermodynamics

At OCV, the anode potential is governed by the equilibrium of the present fuel gas composition, this holds also for operation with hydrocarbon fuels. Over active SOFC anodes, syngas mixtures generated through IIR and even methane/steam mixtures for IIR are supposed to be brought into equilibrium quickly (cf. Section 2.6.1). From this perspective, the electrochemically active sites are within an active catalyst of several hundred μm of thickness, namely the anode support. It is hence safe to assume that the present fuel gas compositions are in equilibrium.

This further implicates that all theoretically possible anode-side half-reactions are in partial equilibrium. The OCV hence yields the same value for every global oxidation reaction. A common and practical choice is to use the global reaction $O_{2,an}/O_{2,cat}$ [15], with the simple result that, independently of the actual anodic charge transfer mechanism,

$$U_{\mathrm{OCV}} = \frac{RT}{2F} \ln \frac{\sqrt{pO_{2,\mathrm{cat}}}}{\sqrt{pO_{2,\mathrm{an}}}}, \tag{2.41}$$

where $pO_{2,\mathrm{cat}}$ and $pO_{2,\mathrm{an}}$ are the equilibrium oxygen partial pressures at cathode and anode, respectively.

Performance and Loss Mechanisms

A number of works investigated the performance of hydrocarbon fueled Ni/YSZ anodes in SOFC single cells. Weber et al. [112] compared C/V characteristics measured at $950\,°C$ for direct internal steam reforming of methane (CH_4 and H_2O as fuel), operation with a simulated reformate (H_2 and CO) as well as pure H_2. No significant difference in cell performance was reported for the three types of operation. A further series of performance measurements with high fuel flow rates and different H_2/CO mixtures (0-100% CO) showed that only for very high CO portions, the cell performance decreased. It was hence assumed that in syngas mixtures the electrochemical oxidation of H_2 is preferred against CO.

Also Jiang et al. [127] have reported similar results. He compared the performance of cells under operation with H_2/CO mixtures and pure H_2. For CO portions below $50\,\%$, the cell performance was similar to that measured under operation with H_2. Interestingly, the cell performance under syngas operation showed much higher values than those measured under operation with H_2/N_2 mixtures. As in Ref. [112], the results were explained with a preference of H_2 against CO for the electrochemical anode reaction. The results were further linked to the WGS reaction (Eq. 2.37), expecting the reaction to convert the CO with the electrochemically oxidized H_2O, thus producing more H_2.

Lin et al. [114] reported EIS measurements, comparing the loss mechanisms of SOFC anodes operated with methane and with hydrogen. The EIS spectra were recorded under a current load of $400\,\mathrm{mA/cm^2}$ at operating temperatures of 600 and 700 °C. In both cases, the reported impedance spectra comprise basically two arcs. The high-frequency impedance arc, which was reportedly associated with the anode, was substantially larger for methane than for hydrogen. The low-frequency arc however did not show any apparent dependence on the fuel type. Interpreting the high-frequency arc as anode activation polarization process implies that the electrochemical fuel oxidation mechanism is substantially different for reformate and hydrogen operation, which is contradictive to the results discussed above.

At this point, it has to be pointed out that a clear interpretation of the recorded spectra would require detailed knowledge of (i) the individual polarization processes that form the measured spectra and of (ii) the operating parameters prevailing during the EIS measurement (cf. Section 2.4.2). In particular, the second aspect is generally complicated by the concentration and temperature gradients, which arise in hydrocarbon fueled SOFC anodes. Any catalytic fuel conversion inhibits measurements under strictly defined gas compositions, an aspect which is essential for the interpretation of electrochemical measurement data (cf. Section 2.5). The same holds for the temperature distribution in real SOFC cells, which experiences strong gradients induced by the strong exothermicity of the reforming reactions, thus inhibiting clear analysis of the thermally activated charge transfer reactions at the electrodes (cf. Section 2.5.2).

A detailed modeling work is presented by Zhu et al. [128], which gives insights on the electrochemical transport mechanisms in hydrocarbon fueled SOFC anodes. Based on the numerical simulation of porous-media transport and detailed heterogeneous reforming chemistry within SOFC anodes, the paper reports simulated EIS spectra for operation with humidified ($3\,\%$ H_2O) methane and hydrogen. While for hydrogen operation, Warburg-type gas transport is clearly reproduced, the model predicts multi-arc impedance responses for hydrocarbon operation. The multiple arcs reveal a complex coupling of gas transport and heterogeneous reforming chemistry.

Despite considerable attention especially toward the catalytic aspects of hydrocarbon conversion, the fundamental processes responsible for electrochemical charge exchange from hydrocarbon fuels are not yet well understood. From the results on Ni/YSZ anodes summarized herein, the preference of H_2 against other electrochemically active species for the anode reaction is suggested, however a clear mechanistic understanding of the charge transfer reaction has not been provided so far. Concerning the reactant gas transport, an interesting coupling of gas diffusion and multi time-scale reforming chemistry is predicted in simulation results. A validation with experimental data has not been provided yet.

Theoretically, EIS is capable to elucidate (i) charge transfer mechanisms and (ii) transport processes possibly coupled with reforming chemistry for hydrocarbon fueled SOFC anodes. This however requires a clear understanding of the individual loss processes that form the inner resistance of the cell as well as analyses under strictly defined operating parameters.

2.6.3. Fuel Impurities

A drawback of currently available fuels for SOFCs is the presence of various contaminants, which have negative impact on performance and long-term stability of the cells.

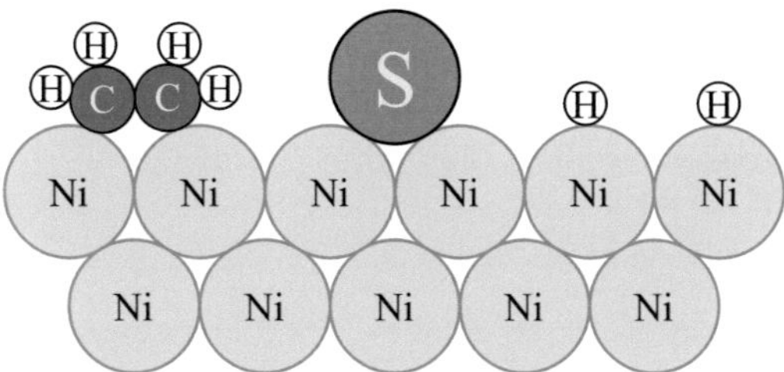

Figure 2.14.: Schematic representation of the poisoning of a catalyst (Ni) surface by sulfur atoms (adapted from Bartholomew [136]).

Sulfur compounds, known to act as a catalyst poison, can be found in nearly all hydrocarbon sources and have hence been paid most attention. Current research equally aims at the understanding of the interactions between sulfur and the electrochemistry of the cell as well as the development of sulfur tolerant anodes. In this subsection, mainly the first aspect is addressed. Comprehensive overviews on both aspects can be found in Refs. [129–131].

Under SOFC relevant conditions, sulfur appears in the form of hydrogen sulfide (H_2S) [101, 132], with concentrations typically in the range of 0.1 to 10 ppm [133]. When H_2S enters the anode compartment of a SOFC, severe performance degradation is experienced, which is generally ascribed to the poisoning of the catalytically active Ni surfaces within the anode.

Sulfur Poisoning Mechanism

The mechanisms of the interactions between H_2S and the surfaces of metal catalysts, including nickel, are well understood [134–136]. H_2S decomposes on the metal surface, forming hydrogen and chemisorbed sulfur in an exothermal reaction. For nickel surfaces, chemisorbed sulfur is one of the most stable adsorption species.

The mechanisms by which chemisorbed sulfur affects the catalytic activity of metal surfaces are manifold. On the basis of the schematic illustration in Fig. 2.14, the effects can be categorized as follows.

- **Reduced amount of active sites:** The adsorbed sulfur blocks several adsorption sites on the catalyst surface. Additionally, the neighboring atoms are electronically modified, thereby affecting their abilities to adsorb and/or dissociate reactant molecules.

- **Retarded surface transport:** The adsorbed sulfur furthermore obstructs transport of the adsorbed reactants on the catalyst surface, by blocking access of the adsorbed reactants to each other.

- **Surface degradation:** A third effect may be the restructuring of the surface by the strongly adsorbed poison, resulting in smoothened surface morphologies. This is possibly causing dramatic changes in catalytic properties, especially for reactions sensitive to surface structure.

While the first and second point represent reversible effects, it has to be pointed out that the third effect might implicate irreversible damaging of the catalyst surface morphology.

Sulfur poisoning in nickel based SOFC anodes hence affects basically any catalytically activated reaction. This means that both charge transfer chemistry (i.e., the electrochemical fuel oxidation at the anode, cf. Eq. 2.3) as well as heterogeneous reforming chemistry (Eqs. 2.36 to 2.38) are possibly affected by sulfur poisoning.

Sulfur Poisoning Studies on SOFC Anodes

A great variety of experiments is reported in SOFC literature, investigating the effects of sulfur poisoning by measuring the degradation of cell performance or cell resistance. The investigations are performed on various SOFC types and anode materials, for a broad range of operating parameters.

Regardless of the investigated material system, there is general consensus on the following dependencies of sulfur poisoning on various parameters, such as

- **Time dependence:** Sulfur poisoning has a characteristic time dependence with two stages: A dramatic (reversible) drop in cell performance followed by a slow but continuous (irreversible) performance degradation [129, 137]. The two stages are most probably related to the above described effects of reversible site blocking and irreversible surface restructuring, respectively.

- **H_2S concentration:** For low H_2S concentrations, the extent of sulfur poisoning increases significantly with increasing H_2S concentration. For H_2S concentrations above $1\dots2$ ppm however, further increase of the H_2S produces only minor increase of the extent of sulfur poisoning [129, 137].

- **Current density:** The extent of sulfur poisoning, both in rapid voltage drop and slow degradation, decreases with increasing load currents [129, 138]. There is evidence that the effect of the slow, irreversible performance degradation due to microstructural changes can be minimized by the application of reasonably high current densities ($1\ \mathrm{Acm}^{-2}$ in Ref. [139]).

- **Operating temperature:** The extent of sulfur poisoning is less pronounced at higher operating temperatures [129, 137, 138]. It is further reported that with increasing operating temperatures, the kinetics of poisoning and regeneration proceed faster [140].

- **Fuel humidification:** It is reported that the extent of sulfur poisoning increases with increasing steam partial pressures at the anode [141]. This indicates that the amount of sulfur adsorbed on the surface is related to the ratio of H_2 and H_2S concentrations at the anode.

Apart from the common assignment of the observed effects to catalyst poisoning, it is worth to mention that several authors assume an impact of sulfur on the ionic conductivity of the electrolyte phase of the MEA. This assumption is based on the fact that cells with scandia stabilized zirconia (SSZ) or GDC as electrolyte and ion conducting phase in the anode show less degradation under exposure to sulfur-containing fuels than cells with YSZ [133, 137, 142]. A detailed analysis by means of EIS would be capable to give insights on the effects of sulfur on ion conduction in electrolyte and cermet anode. In the EIS spectra, a clear impact on the ohmic resistance of the cell R_0 should be observable (cf. Section 2.5.1). For the reported data sets however, no deeper insights could be gained, since the measured spectra were affected by strong inductive contributions, thus inhibiting a clear assignment of the ohmic resistance.

In general, it would be very interesting to analyze the effects of sulfur poisoning on the individual loss contributions of a cell. No detailed EIS analysis of the impact of sulfur on the single polarization processes at SOFC anodes has been reported so far. This might be due to the fact that in the reported poisoning studies, no general understanding of the individual loss contributions to the measured EIS spectra was available.

The impact of sulfur poisoning on the reforming chemistry within SOFC anodes has also been demonstrated experimentally [142]. In particular, it has been shown that the conversion of CO via the WGS reaction within nickel-based SOFC anodes can be seriously affected [143]. At this point, it has to be noted that it is tedious to clearly separate between the impact of sulfur poisoning on the reforming chemistry and on the electrochemistry of the cell. The reforming reactions are strongly exothermic and lowered reaction rates, as observed in the case of sulfur poisoning, hence implicate changes in the cell temperature. Alteration of the cell temperature in turn, might strongly affect the thermally activated electrochemical loss processes (cf. Section 2.5).

3. Experimental

In this chapter, the technical realization of the experiments performed in this work is addressed. A detailed description of the cell design and material composition of the analyzed anode supported single cells is followed by an explanation of the measurement setup used for the electrochemical impedance measurements. The description of the operating parameter range, over which the measurements were carried out, emphasizes the application of model reformate fuels, which guarantee measurements under well-defined fuel gas compositions in this study. Concluding this chapter, the quality of the obtained impedance measurement data is assessed by means of the Kramers-Kronig analysis.

3.1. Samples

The solid oxide fuel cell (SOFC) single cells analyzed in this work are state of the art planar anode supported cells (ASCs) manufactured by Forschungszentrum Jülich (FZJ). The FZJ ASC is one of the best performing, most stable and most reproducible cell SOFC systems available. Figure 3.1 illustrates the experimental layout of the membrane electrode assembly (MEA).

The single cells are built on $50 \times 50 \, \mathrm{mm}^2$ anode substrates (Ni/8YSZ) with an average thickness of about $1000 \, \mu\mathrm{m}$. On these substrates, an anode functional layer (AFL:

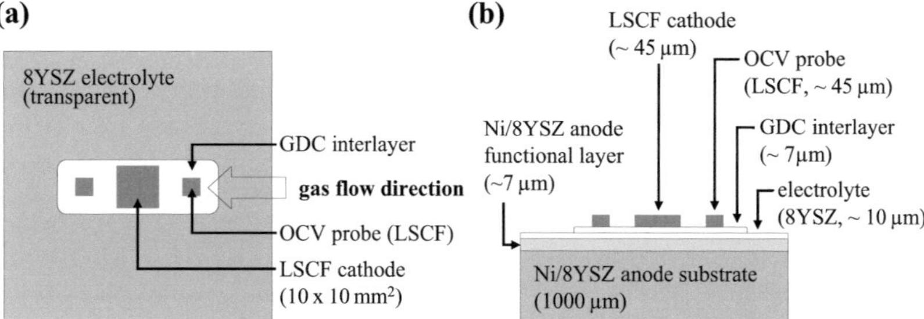

Figure 3.1.: Design of the planar ASC single cells applied for experimental analysis in this thesis: (a) top view and (b) side view (not to scale).

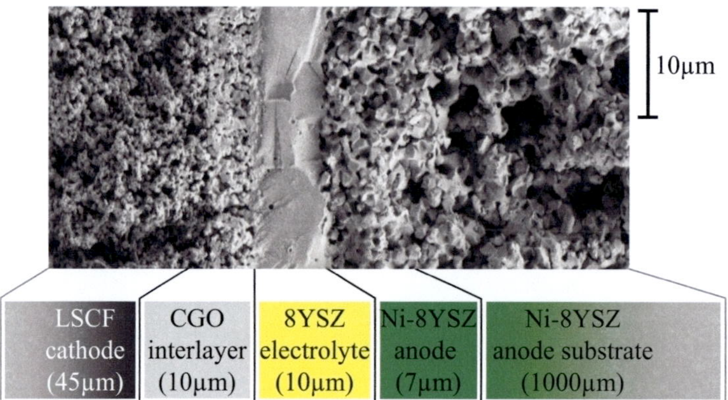

Figure 3.2.: Scanning electron micrograph of a cross section through a fractured FZJ ASC showing the layers of the planar MEA [50].

Ni/8YSZ, approximately $7\,\mu m$ thick) and an electrolyte (8YSZ, approximately $10\,\mu m$ thick) are deposited and co-fired at $1400\,^{\circ}C$. A $Ce_{0.8}Gd_{0.2}O_{2-\delta}$ interlayer is screen-printed and sintered on the electrolyte (approximately $7\,\mu m$ thick) [144]. This gadolinium doped ceria (GDC) interlayer is used to prevent a chemical reaction between LSCF and 8YSZ, which otherwise forms an insulating layer of $SrZrO_3$ [8, 51]. On top of the interlayer a $La_{0.58}Sr_{0.4}Co_{0.2}Fe_{0.8}O_{3-\delta}$ cathode is applied by screen-printing with a thickness of approximately $45\,\mu m$ after sintering. The active cathode area is $10 \times 10\ mm^2$. Two auxiliary electrodes in gas flow direction in front of and behind the cathode are applied for open circuit voltage (OCV) control. The cathodes are separated from the electrolyte by a continuous GDC interlayer with lateral dimensions of $12 \times 30\ mm^2$.

A scanning electron micrograph of the cross section through a fractured cell showing part of the porous anode, the dense electrolyte, the GDC buffer layer, and the porous cathode, is depicted in Fig. 3.2. Details about the cell materials, preparation, microstructure and performance of this cell type are described in detail in Refs. [145–148].

3.2. Test Setup

Figure 3.3 gives an illustration of the measurement setup employed in this work. With the single cells mounted into ceramic Al_2O_3 housings, the cathode is contacted by a gold mesh (1024 meshes/cm^2, $0.06\,mm$ wire thickness). On the anode side, an area of $10 \times 10\ mm^2$ aligned with the cathode is contacted by a nickel mesh (3487 meshes/cm^2, $0.065\,mm$ wire thickness). Accordingly, both electrodes have an active electrode area of $10 \times 10\ mm^2$ in the test assembly. The relatively small active electrode area facilitates cell measurements under homogeneously distributed operating parameters. Gold rings are used for sealing of the gas compartments.

The ceramic housings are placed in a vertical tube furnace, which guarantees defined constant operating temperatures of the cell during the measurements. The cells are operated under ambient pressure with air on the cathode side. Various gas compositions (including H_2, H_2O, CO, CO_2 and N_2) were applied as fuels at the anode side. Gas compositions and flow rates are controlled by a gas mixing and metering unit. High contents of water

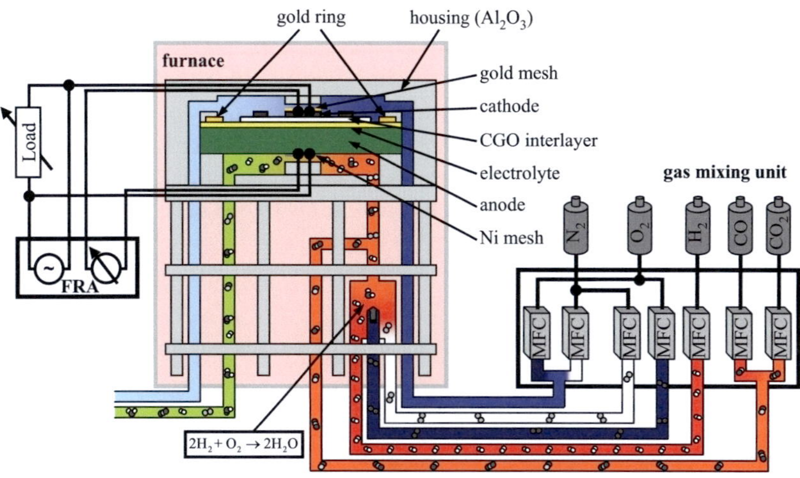

Figure 3.3.: Sketch of the setup designed for SOFC single cell measurements under operation with reformate gases.

vapor in the fuel are realized by feeding O_2 into an upstream combustion chamber. The resulting gas mixtures are transported through flow manifolds adjacent to both electrodes in a co-flow setup. Detailed descriptions of the employed measurement setup, which was developed at the Institut für Werkstoffe der Elektrotechnik (IWE), can be found in Refs. [69, 149–151].

3.3. Cell Measurements

The electrochemical impedance measurements were performed using a Solartron 1260 frequency response analyzer (FRA) within a frequency range from $10\,\text{mHz}$ to $1\,\text{MHz}$. The current stimulus amplitude is chosen to achieve a voltage response not higher than $12\,\text{mV}$. For the cell type investigated, the internal resistance ranged from about $650\,\text{m}\Omega$ at $700\,^\circ\text{C}$ down to $170\,\text{m}\Omega$ at $800\,^\circ\text{C}$. Correspondingly, the amplitude of the excitation current ranged from 17 to $60\,\text{mA rms}$ (technical upper limit of the FRA output current). All impedance measurements performed within this work were carried out under open circuit condition (OCC).

The cells were operated under ambient pressure with air on the cathode side. Three anode-side fuel mixtures were considered, which in this work are referred to as follows:

(i) **hydrogen operation:** Mixtures of H_2, H_2O and N_2.

(ii) **carbon monoxide operation:** Mixtures of CO, CO_2 and N_2.

(iii) **reformate operation:** A syngas mixture of H_2, H_2O, CO, CO_2 and N_2.

The anode and cathode gas flow rates were maintained at a constant value of $250\,\text{sccm}$ during all experiments. The experimental measurements were taken at operating temperatures between 680 and $800\,^\circ\text{C}$ and under a broad range of H/C/O-ratios of the anode fuel.

Due to the high operating temperatures and the catalytic activity of the Ni surfaces within the anode substrate, reformate fuels of arbitrary composition are converted within the anode until chemical equilibrium is reached. The gas composition at chemical equilibrium depends on the temperature and the elementary composition (i.e., the H/C/O ratio) of the gas mixture. In order to perform measurements under well-defined gas compositions, the equilibrium gas compositions for each measurement point were calculated with the thermodynamic database MALT [152] in this work. Then the fuel was introduced to the anode in its calculated equilibrium composition. Carbon deposition within the anode was prevented in all measurement points. For that reason, all the applied fuel gas compositions were chosen in a way that the share of coke in the corresponding equilibrium composition was zero (cf. Appendix B).

Due to inevitable errors of the mass flow controllers (MFCs) and of gas leakage in the measurement setup, an adjustment of the gas supply had to be carried out:

- In the first step, the share of CO and CO_2 in the equilibrium reformate mixture was replaced by the same amount of N_2. The equilibrium concentration for this reference fuel composition was calculated. With the equilibrium partial pressures of oxygen at the anode ($pO_{2,\mathrm{an}}$) and at the cathode ($pO_{2,\mathrm{cat}}$), the theoretical OCV for operation with the applied fuel was calculated via the Nernst equation (Eq. 2.41). The reference fuel was introduced to the anode and the exact H_2/H_2O ratio was adjusted until the measured OCV reached the theoretical value expected in case of chemical equilibrium. After the adjustment, an initial reference impedance measurement under hydrogen operation was conducted.

- In the second step, immediately after the initial measurement, the N_2 in the fuel gas was replaced by CO and CO_2. The ratio of CO and CO_2 was adjusted until the OCV reached the corresponding theoretical value for the equilibrium reformate composition. Subsequently, the proper impedance measurement under reformate operation was performed.

The presented measurement procedure has the advantage that the exact gas composition within the anode is known for every measuring point. Hence a highly accurate characterization of the electrochemical impedance spectra, which exhibit a notable sensitivity on the gas compositions at the electrodes, is guaranteed. Additionally, reference impedance measurements under hydrogen operation were realized via this procedure. The partial pressures of H_2 and H_2O were the same as for reformate operation and the reference measurements were conducted immediately before the reformate measurements. Hence, the recorded impedance spectra can be compared for both operation modes under identical $pH_{2,\mathrm{an}}$ and pH_2O_{an}.

3.4. Measurement Data Quality

For a successful interpretation of impedance spectra, the measurement data quality is of crucial importance. The quality and amount of information that can be extracted from impedance data is implicitly connected to the noise-level and the compliance of the measured curve with the above mentioned principles of causality, linearity and stability (cf. Section 2.4.1).

A well-established method used to asses the consistency and quality of measured impedance spectra is the Kramers-Kronig validation [63]. The Kramers-Kronig relations are integral equations, which constrain the real and imaginary components of the impedance for systems that satisfy the conditions of causality, linearity, and stability [63].

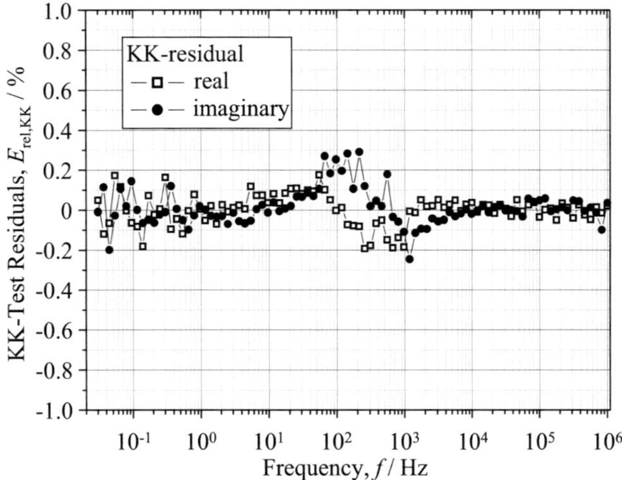

Figure 3.4.: Residual pattern of the Kramers-Kronig test performed on a typical impedance spectrum recorded for reformate operation. [cell# Z2_190]

In this work the necessary compliance of the measured data with the Kramers-Kronig transformation rules is verified by using the KK test for Windows software [153, 154]. Figure 3.4 shows a Kramers-Kronig validation for a typical impedance spectrum. For most part of the spectrum the relative errors of both real and imaginary data lie below a very low value of 0.4 %.

4. Process Identification

In this chapter, the electrochemical processes, which determine the performance of reformate fueled anode supported solid oxide fuel cells (SOFCs) are identified by means of Electrochemical Impedance Spectroscopy (EIS). The applied approach via EIS measurements on single cells and the subsequent analysis via the distribution of relaxation times (DRT) method allows for the deconvolution of the single polarization processes from the measured spectra.

The underlying physical loss processes are identified through a detailed characterization under a broad variation of operating parameters. Based on this knowledge, a physically motivated equivalent circuit model (ECM) of the anode supported cells (ASCs) investigated herein is developed and presented in the end of this chapter. With the help of this ECM, the measured impedance spectra can be separated into the single loss contributions, thus permitting a precise determination of the area specific resistance (ASR) for the single loss contributions of the SOFC single cells. Parts of the results presented in this chapter have been published in Ref. [155].

4.1. Comparison to Hydrogen Operation

In order to understand the polarization processes that determine the electrochemistry of reformate fueled ASCs, the well-understood electrochemical polarization processes occurring under hydrogen operation (cf. Chapter 2) were used as a reference. Figure 4.1 gives the comparison of typical electrochemical impedance spectra measured under hydrogen and reformate operation. Table 4.1 gives the detailed gas compositions of the fuel. The cathode was operated with air and the cell temperature was $800\,°C$.

As can be expected, the ohmic resistance of the cell (R_0) is not depending on the anode fuel. Accordingly, the high-frequency intersections of the measured spectra with the real axis are identical for both operation types. However, the polarization resistances R_{pol} and particularly the shape of the measured EIS spectra are different. Whereas the high-frequency arcs are similar in shape and characteristic frequency (ca. $10\,\mathrm{kHz}$), the low-frequency section of the measured impedance spectra are considerably different for the different fuels. For hydrogen operation, the spectrum is dominated by one single arc with characteristic frequencies around $20\,\mathrm{Hz}$. For reformate operation, the low-frequency

Table 4.1.: Anodic partial pressures of H_2, H_2O, CO and CO_2 selected for the comparison of EIS measurements under hydrogen and reformate operation.

operation mode	$pH_{2,an}$ (atm)	pH_2O_{an} (atm)	pCO_{an} (atm)	$pCO_{2,an}$ (atm)
hydrogen	0.70	0.30	0	0
reformate	0.35	0.15	0.35	0.15

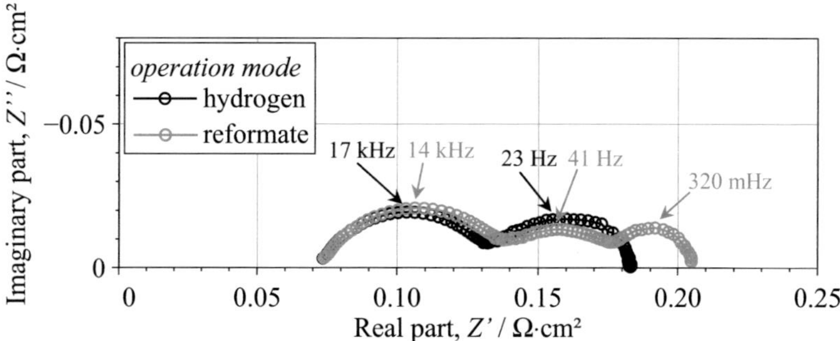

Figure 4.1.: EIS spectra measured under hydrogen and reformate operation (fuel compositions see Table 4.1; cathode: air; $T = 800\,°C$). [cell# Z2_190]

section of the measured impedance spectra is characterized by two distinctive impedance arcs. Additionally to the semicircle at ca. 40 Hz, an additional semicircle appears at characteristic frequencies below 1 Hz.

The DRT calculated from the measured spectra, as depicted in Fig. 4.2, gives more insight on the single polarization processes that form the measured impedance arcs. As for hydrogen operation, the two high-frequency peaks originating from the anode activation polarization (P_{2A} and P_{3A}) are observed when the cell is operated on reformate. It is interesting to see that size, shape, and characteristic relaxation frequencies of the peaks are very similar for both operation modes. The cathode activation polarization process (P_{2C}) is not dependent on the fuel variation. The corresponding peak appears at intermediate frequencies (ca. $50\dots100$ Hz) and is visible for both operation modes. The peak of P_{2C} is overlapping with the peak of the anode concentration polarization process (P_{1A}, characteristic frequencies around 10 Hz) [57, 86], which is more prominent for hydrogen operation. When the cell is operated with reformate, this process is much smaller than for hydrogen operation, but it is still clearly visible.[1] At frequencies below 1 Hz, an additional polarization process is visible. This process is originating the additional low-frequency semicircle in the measured spectra. Due to its occurrence under reformate operation, it is probably linked to the reforming reactions within the anode substrate. It is called P_{ref} in this work.

Apparently, the polarization processes, which have been identified and characterized un-

---

[1]Both GFLW and Gerischer impedance exhibit asymmetric DRTs, characterized by a large peak located at the characteristic frequency followed by smaller peaks at higher frequencies [7]. For reformate operation with a smaller P_{1A}, the second peak of P_{2C} becomes visible.

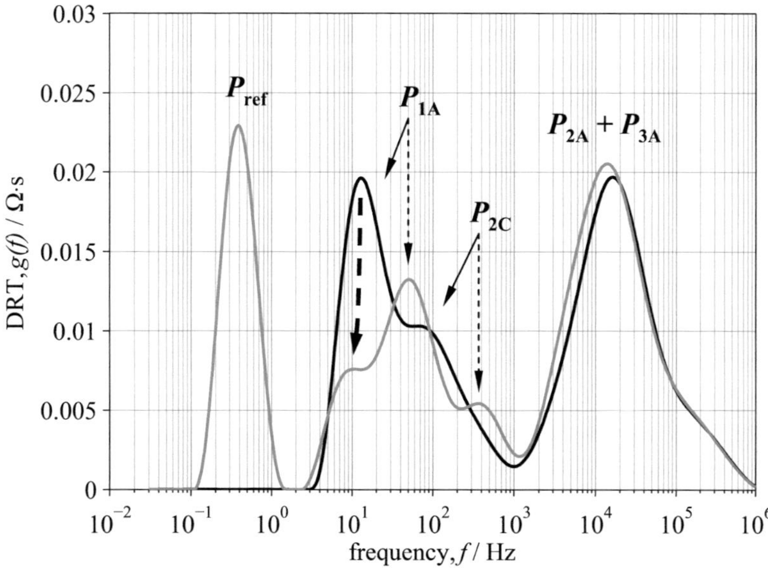

Figure 4.2.: DRT calculated from the measured EIS spectra given in Fig. 4.1. The black line denotes hydrogen operation and the gray line denotes reformate operation. [cell# Z2_190]

der operation with hydrogen as fuel and air as oxidant (P_{1A}, P_{2C}, P_{2A} and P_{3A}) [86] can be identified in the EIS spectra recorded for operation with reformate fuels, as well. Additionally, a polarization process never reported before is identified at low frequencies (P_{ref}), which is probably related to the occurrence of the reforming reactions at the anode.

In order to (i) prove the performed identification of the well-known polarization processes for reformate operation and (ii) investigate the physical origin of the newly identified polarization process P_{ref}, series of EIS measurements under systematic variation of operating parameters (fuel humidification, temperature and gas composition of the fuel) were carried out. In the following sections, the single polarization processes are analyzed with respect to their dependence on the varied operating parameters.

4.2. Impact of Fuel Humidification

In order to confirm the assignment of the identified polarization processes to their occurrence at the anode, EIS measurements for varied fuel humidification were performed. The humidification of the model reformate fuel (75 % H_2 and 25 % CO) was varied stepwise between 20 and 60 %. The detailed gas composition of the fuel is given in Table 4.2. Figure 4.3 depicts the recorded impedance spectra and the corresponding DRT.

The assigned anodic polarization processes P_{ref}, P_{1A}, P_{2A} and P_{3A} exhibit a clear dependence on the humidification of the reformate fuel. The anode activation polarization processes exhibit the same qualitative behavior as reported for hydrogen operation [7,86]. The decrease in fuel humidification causes a strong increase of P_{2A} and a minor increase of P_{3A}. The process assigned to the anode diffusion polarization (P_{1A}) exhibits a moderate dependence on the parameter variation, whereas the newly identified process P_{ref} is

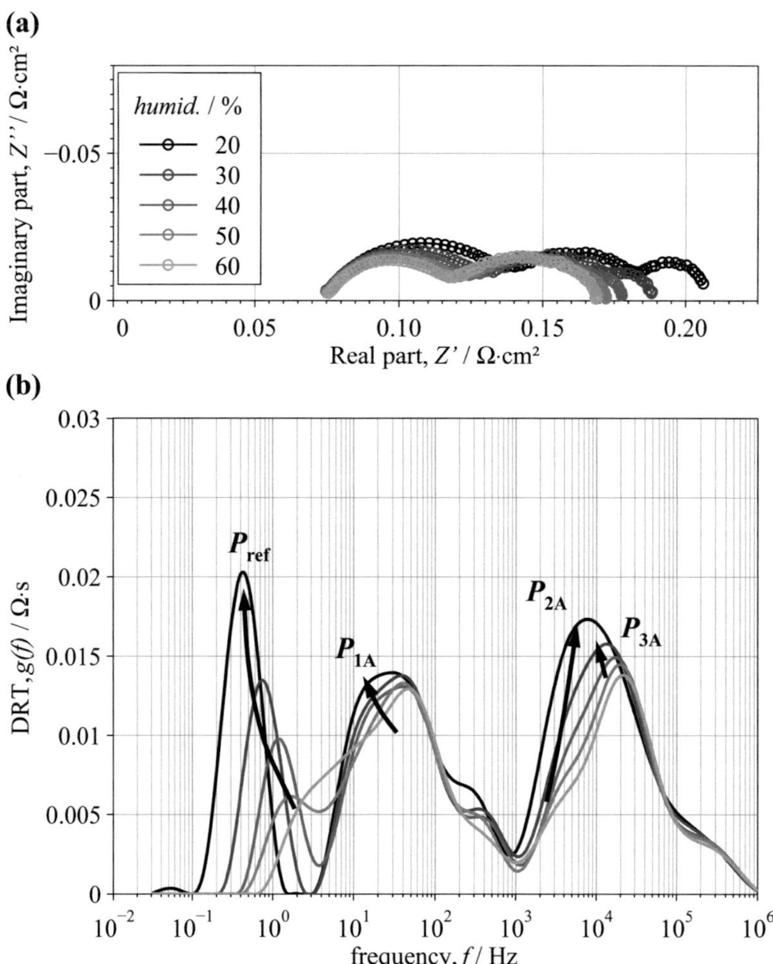

Figure 4.3.: (a) Series of impedance spectra and (b) corresponding DRT recorded for varying humidification of the anode fuel gas (fuel composition see Table 4.2; cathode: air; $T = 800\,^{\circ}\mathrm{C}$). [cell# Z2_188]

Table 4.2.: Anodic partial pressures of H_2, H_2O, CO and CO_2 selected for the EIS measurements under variation of the fuel humidification.

humidification (%)	$pH_{2,an}$ (atm)	pH_2O_{an} (atm)	pCO_{an} (atm)	$pCO_{2,an}$ (atm)
20	0.60	0.15	0.20	0.05
30	0.53	0.22	0.17	0.08
40	0.45	0.30	0.15	0.10
50	0.38	0.37	0.12	0.13
60	0.30	0.45	0.10	0.15

significantly affected. With decreasing fuel humidification, this process exhibits a strong increase in ASR along with a decrease of the characteristic relaxation frequency.

4.3. Impact of Operating Temperature

The dependence on the operating temperature distinguishes diffusion polarization processes with low temperature dependence from thermally activated activation polarization processes (cf. Chapter 2). In order to analyze the temperature dependence of the identified electrochemical processes at the anode, impedance measurements were performed for varying operating temperatures. Therefore, the cell temperature was varied between 680 and 800 °C in steps of 30 K. The applied equilibrium fuel gas compositions are given in Table 4.3.[2] Figure 4.4 depicts the impedance spectra recorded for this temperature variation along with the corresponding DRT plots.

Table 4.3.: Anodic partial pressures of H_2, H_2O, CO, CO_2 and N_2 selected for the EIS measurements under variation of the operating temperature T.

T (°C)	$pH_{2,an}$ (atm)	pH_2O_{an} (atm)	pCO_{an} (atm)	$pCO_{2,an}$ (atm)
800	0.25	0.25	0.25	0.25
770	0.26	0.24	0.24	0.26
740	0.27	0.23	0.23	0.27
710	0.28	0.22	0.22	0.28
680	0.28	0.22	0.21	0.29

The strong temperature dependence of the anodic activation polarization processes P_{2A} and P_{3A}, which indicates the expected thermal activation behavior, can be seen immediately. Along with the significant increase in resistance, the characteristic frequencies are decreasing with decreasing operating temperature. Also thermal activation of the process P_{2C} (activation polarization of the cathode) can be observed, which at lower temperatures

---

[2]According to the thermodynamic equilibrium of the water-gas shift (WGS)-reaction, the equilibrium composition shifts slightly toward H_2/CO_2 for lower temperatures (cf. Section 2.6.1).

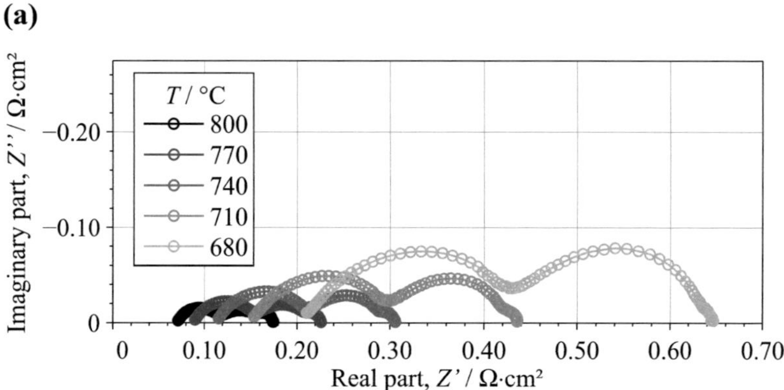

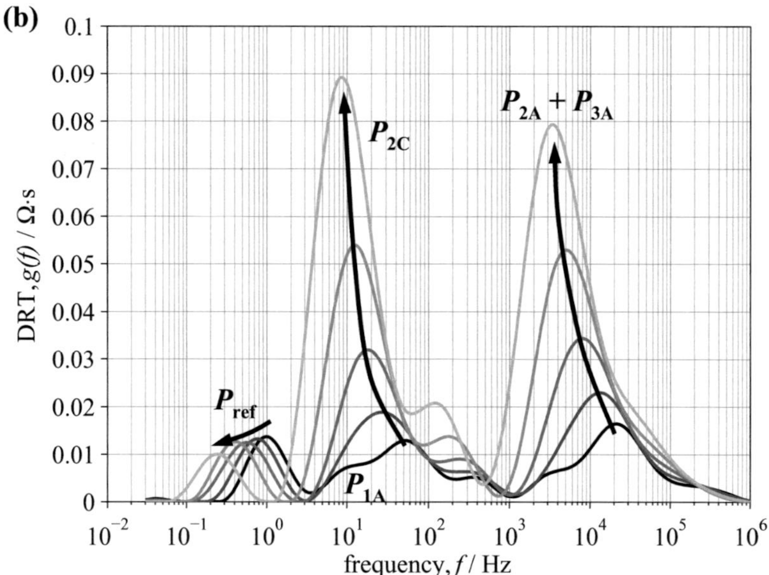

Figure 4.4.: (a) Series of impedance spectra and (b) corresponding DRT recorded for
varying operating temperature (fuel composition see Table 4.3; cathode: air).
[cell# Z2_190]

completely covers P_{1A}. As can be expected for a diffusion polarization process, P_{1A} exhibits a negligible dependence on the variation of temperature. It is interesting to see that the newly identified low-frequency process P_{ref}, nearly unaffected in resistance, is shifting to lower characteristic frequencies with decreasing operating temperatures.

4.4. Impact of Reformate Composition

In order to investigate the influence of the reformate fuel composition on the identified anodic polarization processes, EIS spectra have been recorded for varying shares of H_2/H_2O and CO/CO_2 in the anode fuel gas. For this series of measurements, gas compositions corresponding to a fuel humidification of 50% (i.e., $pH_2O_{an} + pCO_{2,an} = 0.5\,\mathrm{atm}$) were chosen. Starting from a fuel gas composition of 50% H_2/H_2O and 50% CO/CO_2, the share of CO/CO_2 was increased stepwise up to 100%. The detailed fuel compositions are given in Table 4.4.

Table 4.4.: Anodic partial pressures of H_2, H_2O, CO, CO_2 and N_2 selected for the EIS measurements for varying shares of H_2/H_2O and CO/CO_2 in the anode fuel gas.

CO/CO_2-share (%)	$pH_{2,an}$ (atm)	pH_2O_{an} (atm)	pCO_{an} (atm)	$pCO_{2,an}$ (atm)
50	0.25	0.25	0.25	0.25
60	0.20	0.20	0.30	0.30
70	0.15	0.15	0.35	0.35
80	0.10	0.10	0.40	0.40
90	0.05	0.05	0.45	0.45
100	0.00	0.00	0.50	0.50

Figure 4.5 illustrates the measured spectra along with the corresponding DRT for different shares of CO/CO_2 in the reformate fuel. For the measurements with 50 to 90% CO/CO_2 in the fuel, the polarization processes attributed to the anode activation polarization ($P_{2A} + P_{3A}$) exhibit relatively low dependence on the variation of the fuel gas composition. Only at the last step, when the cell is operated with pure CO/CO_2, the DRT shows a dramatic increase of the polarization processes P_{2A} and P_{3A}.

The anode activation polarization processes hence exhibit a significant dependence on the presence of H_2/H_2O in the fuel. When the anode is operated with pure CO/CO_2, the anode activation polarization processes exhibit a significant higher ASR to when H_2/H_2O is present. This behavior must be ascribed to the electrochemical fuel oxidation mechanism, which will be investigated in Chapter 5.

In contrast to P_{2A} and P_{3A}, the anodic low-frequency polarization processes ($P_{1A} + P_{ref}$) exhibit continuous shifts for the variation of the fuel gas composition. As the share of CO/CO_2 increases, P_{1A} decreases, whereas P_{ref} increases continuously. For operation with pure CO/CO_2, P_{1A} vanishes, while the remaining P_{ref} must represent the transport of CO/CO_2 within the gas pores of the anode substrate [60].

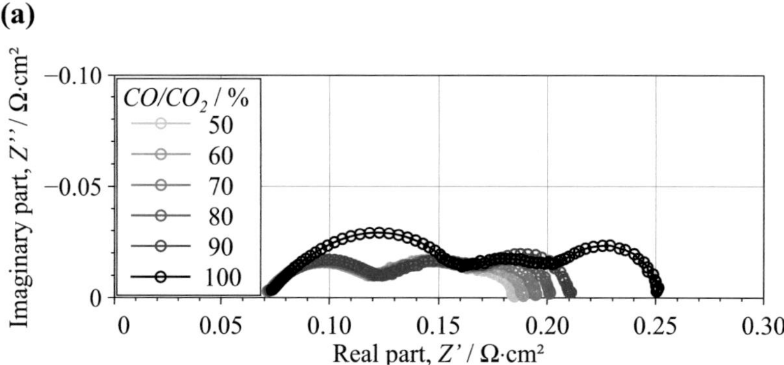

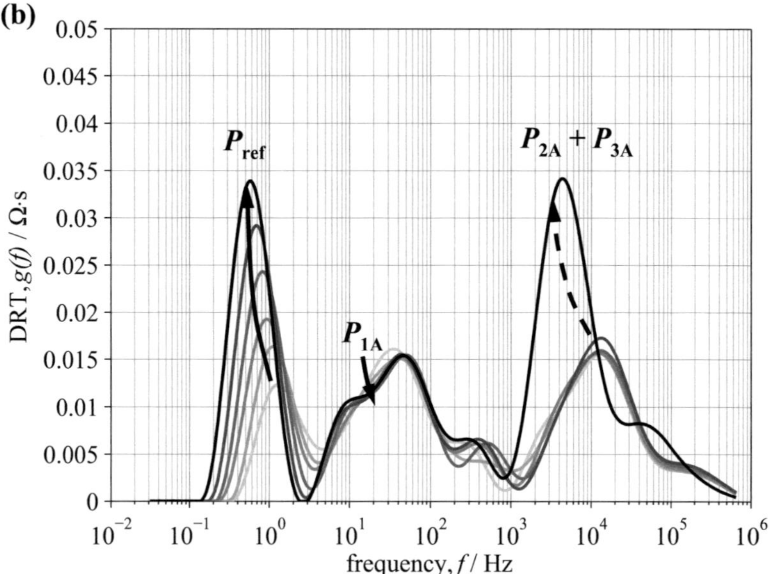

Figure 4.5.: (a) Series of impedance spectra and (b) corresponding DRT recorded for varying shares of H_2/H_2O and CO/CO_2 in the anode fuel gas (fuel composition see
Table 4.4; cathode: air; $T = 800\,°C$). [cell# Z2_211]

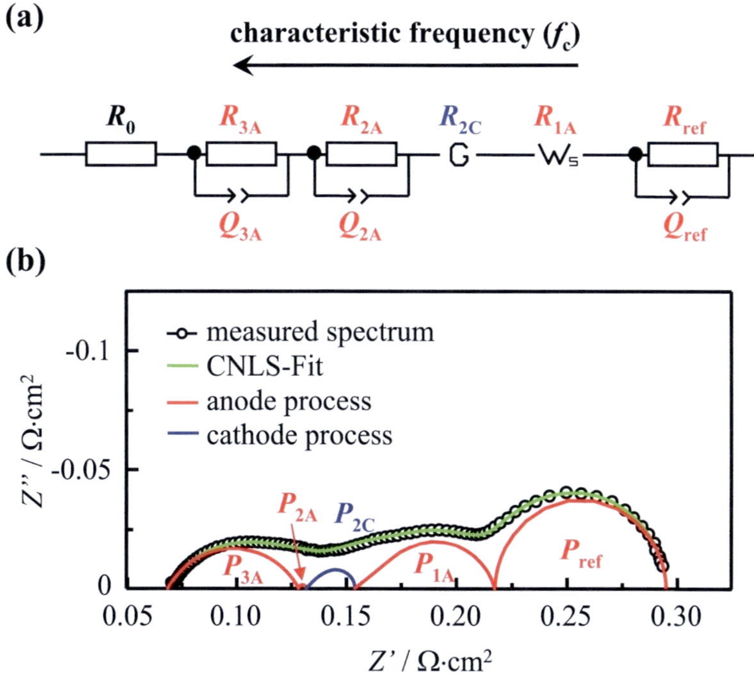

Figure 4.6.: (a) Extended equivalent circuit model used for the CNLS-fit of the impedance data measured under reformate operation. (b) CNLS-fit of an impedance spectrum measured under $pH_{2,an} = 0.15$ atm, $pH_2O_{an} = 0.05$ atm, $pCO_{an} = 0.15$ atm, $pCO_{2,an} = 0.05$ atm, balance: N_2; cathode: air; $T = 800\,°C$. [cell# Z2_190]

The observed behavior indicates that for reformate fueled ASCs, the low-frequency polarization process P_{1A} has to be attributed to the diffusion of H_2/H_2O, whereas P_{ref} is dominated by the diffusion of CO/CO_2. During the whole fuel variation, the transformation of P_{1A} and P_{ref} proceeds continuously — even in the last step, where the gas composition switches to $100\,\%$ CO/CO_2. In terms of reforming chemistry, the pure CO/CO_2 fuel gas mixture is inert, i.e., no reforming reactions can occur. This indicates that the reforming reactions, which theoretically might be excitated during the EIS measurement, are not appearing as a single polarization process. In Chapter 6, the interactions between reforming chemistry and the process P_{ref} will be analyzed in more detail.

4.5. Equivalent Circuit Model Definition

The extensive DRT analysis presented in the preceding sections has confirmed the identification of the single polarization processes, which form the EIS spectra measured on anode supported SOFCs operated on reformate fuels. Based on this knowledge, a physically motivated equivalent circuit model (ECM) representing the impedance of the applied cells under reformate operation is set up. The ECM is depicted in Fig. 4.6a.

All the polarization processes which occur under hydrogen operation have been identified for reformate operation, too. In order to model these processes, the following serial

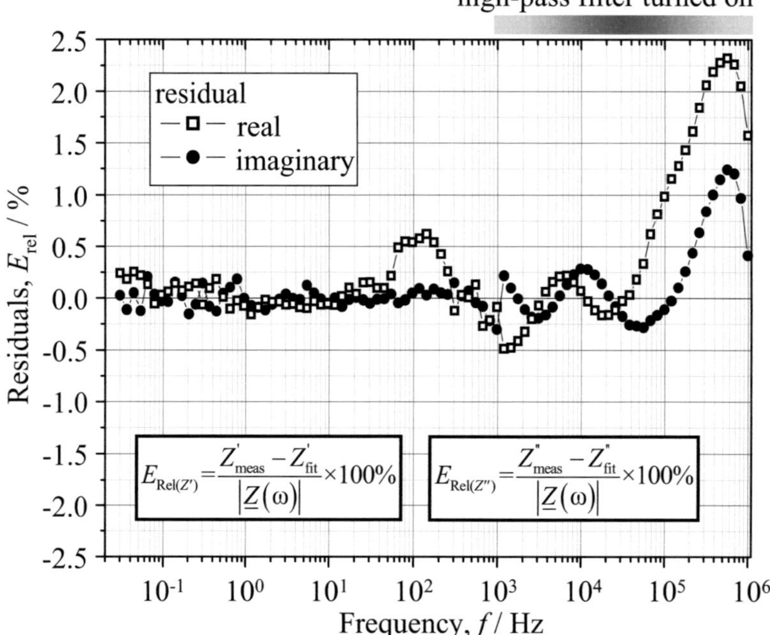

Figure 4.7.: Residual pattern of the CNLS-fit depicted in Fig. 4.6 showing the relative errors E_{rel} of real (Z') and imaginary part (Z'') of the simulated impedance over the investigated frequency range. [cell# Z2_190]

elements were adopted from the ECM proposed by Leonide et al. [86], which has been developed and validated for hydrogen operation: (i) an ohmic resistor (R_0: overall ohmic losses), (ii) two serial RQ elements (P_{2A} and P_{3A}: gas diffusion coupled with charge transfer reaction and ionic transport within the anode functional layer [87]), (iii) a Gerischer element (P_{2C}: activation polarization of the LSCF-cathode [156]), and (iv) a generalized finite length Warburg element (G-FLWS) (P_{1A}: diffusion polarization in the anode substrate). The newly discovered process P_{ref} can be described precisely by an additional RQ element.

Figure 4.6b shows a typical CNLS-fit of the ECM applied to an impedance curve measured under reformate operation. Generally, the residuals (relative errors) are distributed uniformly around the frequency axis, showing no systematic deviation (Fig. 4.7). For most of the spectrum, the relative errors lie below an absolute value of $0.5\,\%$, thus demonstrating the validity of the model. From about $100\,\text{kHz}$ upward, inductive contributions caused by the electrical wiring become noticeable. However, the error still does not exceed a maximum value of $2.3\,\%$ in this frequency range.

With the help of the developed physically motivated ECM and the numerical accuracy of the CNLS-fit, an accurate quantitative analysis of the individual loss contributions for reformate operation of the investigated cells is possible.

4.6. Conclusions

The electrochemical impedance spectra of anode supported SOFCs were characterized for operation with reformate gas as a fuel. Aim of the study presented in this chapter was the identification of the performance-limiting polarization processes of the cells for reformate operation. The characterization was performed over a broad range of operating temperatures and H/C/O-ratios of the anode fuel.

Calculating the DRT enabled the separation of the individual electrochemical processes at the anode. The processes studied under hydrogen operation [7, 86] were found under reformate operation, too: (i) The gas diffusion polarization in the anode substrate and (ii) the coupling of anodic charge transfer reaction and the ionic conduction in the cermet anode (anode activation polarization).

Furthermore, an additional electrochemical process never reported before, with a characteristic relaxation frequency below 1 Hz was identified to occur under reformate operation. The observed parameter dependencies indicate an assignment of this polarization process to the diffusion of CO/CO_2, probably related to the reforming reactions.

Based on this knowledge, a physically motivated ECM was set up, representing the single polarization processes of the analyzed cells operated with reformate fuels. A CNLS-fit procedure of this model to measured impedance spectra permits a precise determination of the ASRs of the single loss contributions to the overall cell resistance. It is the foundation for the quantitative analyses presented in the following chapters.

5. Electrochemical Fuel Oxidation

In this chapter, the electrochemical fuel oxidation mechanism for reformate fueled solid oxide fuel cell (SOFC) anodes is elucidated with the help of Electrochemical Impedance Spectroscopy (EIS). The methodology of this work allows for a quantitative analysis of the anode activation polarization impedance, performed under a systematic variation of characteristic operating parameters. The analysis further leads to the determination of kinetic parameters, from which the mechanism of the electrochemical fuel oxidation is identified unambiguously. The results presented in this chapter have been published in Ref. [155].

5.1. Parameter Study

Starting point of the performed analysis is the assumption that only H_2 is electrochemically oxidized in reformate fueled SOFC anodes. Also in literature, a certain preference of H_2 against other electrochemically active species in hydrocarbon fuels has been assumed several times (see Section 2.6 for details). For the reformate fuels applied herein, the preference of H_2 against CO seems further reasonable, when considering the notably lower activation energy ($E_{\mathrm{act},H_2} = 1.10\,\mathrm{eV}$ [59], $E_{\mathrm{act},CO} = 1.23\,\mathrm{eV}$ [60]).

In Ref. [59], it is shown that the electrochemical oxidation of H_2 in SOFC anodes is uniquely characterized by its dependence on (i) the partial pressure of hydrogen ($pH_{2,\mathrm{an}}$), (ii) the partial pressure of steam (pH_2O_{an}) and (iii) the operating temperature (T). A comprehensive analysis of these characteristic dependencies requires an exact knowledge of the corresponding operating parameters. In earlier studies, the catalytic gas conversion of the applied reformate fuels within the anode support led to uncertainty about the actual gas composition as well as the actual temperature at the electrochemically active sites within the reformate fueled anode (see Section 2.6 for details).

The herein applied method of operation with model reformate fuels (cf. Section 3.3) has the advantage that the EIS measurements can be performed under well-defined gas compositions and operating temperatures. Additionally, reference impedance measurements under hydrogen operation were realized via this procedure. The measurements performed under reformate operation can be compared to measurements under hydrogen operation

Table 5.1.: Anodic partial pressures of H_2, H_2O, CO, CO_2 and N_2 selected for the EIS measurements under variation of $pH_{2,an}$ ($= pCO_{an}$).

operation mode	$pH_{2,an}$ (atm)	pH_2O_{an} (atm)	pCO_{an} (atm)	$pCO_{2,an}$ (atm)	$pN_{2,an}$ (atm)
reformate	0.05	0.10	0.05	0.10	0.70
	0.10	0.10	0.10	0.10	0.60
	0.20	0.10	0.20	0.10	0.40
	0.40	0.10	0.40	0.10	0.00
hydrogen	0.05	0.10	0.00	0.00	0.85
	0.10	0.10	0.00	0.00	0.80
	0.20	0.10	0.00	0.00	0.70
	0.40	0.10	0.00	0.00	0.50

recorded at identical $pH_{2,an}$, pH_2O_{an} and T. The developed physically motivated equivalent circuit (cf. Section 4.5) facilitates a precise determination of the anode activation polarization processes P_{2A} and P_{3A}, which are incorporating the electrochemical fuel oxidation within the anode. Hence, a highly accurate quantitative characterization of the anode activation polarization processes in dependence on the characteristic operating parameters ($pH_{2,an}$, pH_2O_{an}, T) is guaranteed.

The impedance measurements under varied H_2 and H_2O concentrations in the fuel were conducted at $800\,°C$. The partial pressures $pH_{2,an}$ and pH_2O_{an} were varied independently of each other, which means that for the variation of $pH_{2,an}$, the H_2O concentration was fixed and vice versa. Balancing each variation by inert N_2 helped to maintain the anode gas flow rate at a value of $250\,sccm$. It is noted that due to the chemical equilibrium of the model reformate, a variation of $pH_{2,an}$ implies a variation of pCO_{an}. A variation of pH_2O_{an} in analogy implies a variation of $pCO_{2,an}$. According to the chemical equilibrium at $800\,°C$, both the partial pressures of H_2 and CO and the partial pressures of H_2O and CO_2 are in the same range, respectively. Table 5.1 gives the detailed partial pressures for the variation of $pH_{2,an}$ (pCO_{an}) and table 5.1 for the variation of pH_2O_{an} ($pCO_{2,an}$).

The variation of the operating temperature started from $800\,°C$. Carbon deposition within the anode had to be prevented in all measurement points. For that reason, the gas composition was chosen in a way that the share of coke in the equilibrium composition was zero at any point (C-H-O Gibbs diagram are given in Appendix B). The initial concentration ratio of the reformate gas in equilibrium of $pH_2 / pH_2O / pCO / pCO_2 = 0.25 / 0.25 / 0.25 / 0.25$ allowed a temperature variation until $680\,°C$ without reaching the carbon deposition regime. According to the chemical equilibrium, this ratio changed slightly with lower temperatures. The detailed fuel compositions for each temperature point are listed in Table 5.1.

5.1.1. H_2 Partial Pressure Dependence

In the first step, the characteristic dependence of the polarization processes P_{2A} and P_{3A} on the partial pressure of hydrogen, $pH_{2,an}$, was analyzed. EIS measurements were performed under varying H_2 concentration within the reformate fuel. The detailed gas composition for this series of measurements are given in Table 5.1. The H_2 concentration in

Table 5.2.: Anodic partial pressures of H_2, H_2O, CO, CO_2 and N_2 selected for the EIS measurements under variation of pH_2O_{an} ($= pCO_{2,an}$).

operation mode	$pH_{2,an}$ (atm)	pH_2O_{an} (atm)	pCO_{an} (atm)	$pCO_{2,an}$ (atm)	$pN_{2,an}$ (atm)
reformate	0.15	0.05	0.15	0.05	0.60
	0.15	0.07	0.15	0.08	0.55
	0.15	0.12	0.15	0.13	0.45
	0.15	0.17	0.15	0.18	0.35
	0.15	0.25	0.15	0.25	0.20
	0.15	0.35	0.15	0.35	0.00
hydrogen	0.15	0.05	0.00	0.00	0.80
	0.15	0.07	0.00	0.00	0.78
	0.15	0.12	0.00	0.00	0.73
	0.15	0.17	0.00	0.00	0.68
	0.15	0.25	0.00	0.00	0.60
	0.15	0.35	0.00	0.00	0.50

Table 5.3.: Anodic partial pressures of H_2, H_2O, CO, CO_2 and N_2 selected for the EIS measurements under variation of the operating temperature T.

operation mode	T (°C)	$pH_{2,an}$ (atm)	pH_2O_{an} (atm)	pCO_{an} (atm)	$pCO_{2,an}$ (atm)	$pN_{2,an}$ (atm)
reformate	800	0.25	0.25	0.25	0.25	0.00
	770	0.26	0.24	0.24	0.26	0.00
	740	0.27	0.23	0.23	0.27	0.00
	710	0.28	0.22	0.22	0.28	0.00
	680	0.28	0.22	0.21	0.29	0.00
hydrogen	800	0.25	0.25	0.00	0.00	0.50
	770	0.26	0.24	0.00	0.00	0.50
	740	0.27	0.23	0.00	0.00	0.50
	710	0.28	0.22	0.00	0.00	0.50
	680	0.28	0.22	0.00	0.00	0.50

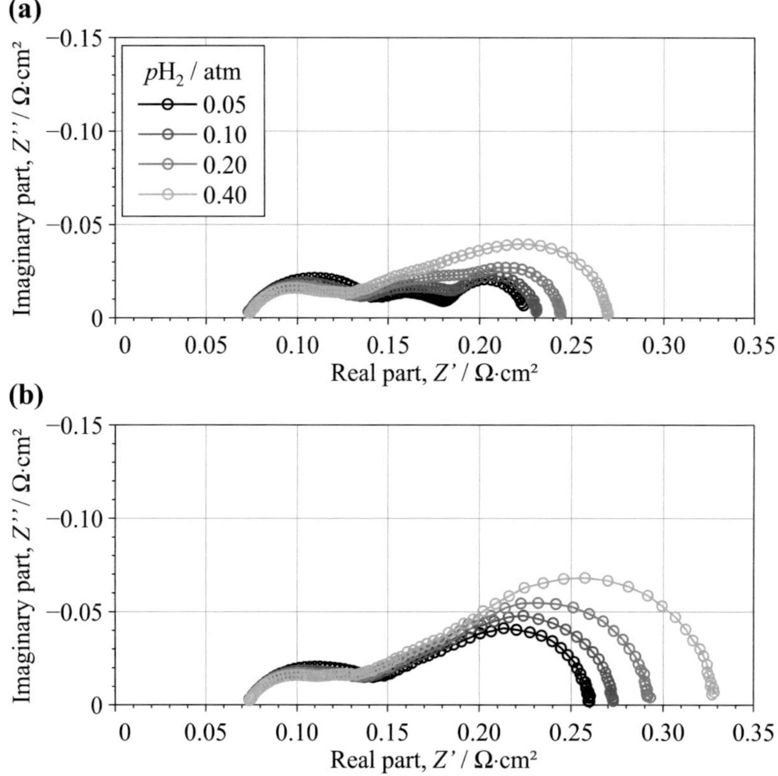

Figure 5.1.: Nyquist representation of measured EIS spectra for varying $pH_{2,an}$. (a) Reformate and (b) hydrogen operation (fuel compositions see Table 5.1; cathode: air; $T = 800\,°C$). [cell# Z2_190]

the fuel was varied in four steps between 5 and 40 %. During the variation of $pH_{2,an}$, the H_2O concentration was fixed as 10 %. According to the chemical equilibrium of the chosen reformate gas mixture at the anode, the CO content was varied in the same way as H_2, whereas the CO_2 content was kept constant. The recorded impedance spectra are depicted in Fig. 5.1.

Figure 5.1a depicts the EIS spectra measured under reformate operation and Fig. 5.1b gives the spectra of the corresponding reference measurements recorded under hydrogen operation. Areas of interest in the measured spectra are the high frequency arcs at the left part of the spectra. These impedance arcs are originated by the anode activation polarization, which is directly linked to the electrochemical fuel oxidation within the anode (cf. Section 2.4). In Fig. 5.1a, it can be observed that for increasing H_2 concentrations, the high-frequency arc becomes slightly greater, indicating a slight increase in anode activation polarization resistance. The high-frequency arcs of the reference spectra measured under hydrogen operation (Fig. 5.1b) exhibit the same characteristic trend for the investigated $pH_{2,an}$ variation. Also shapes and sizes of the arcs seem comparable for both operation types. The corresponding distribution of relaxation times (DRT) calculated from the measured spectra, as depicted in Fig. 5.2, confirm these findings.

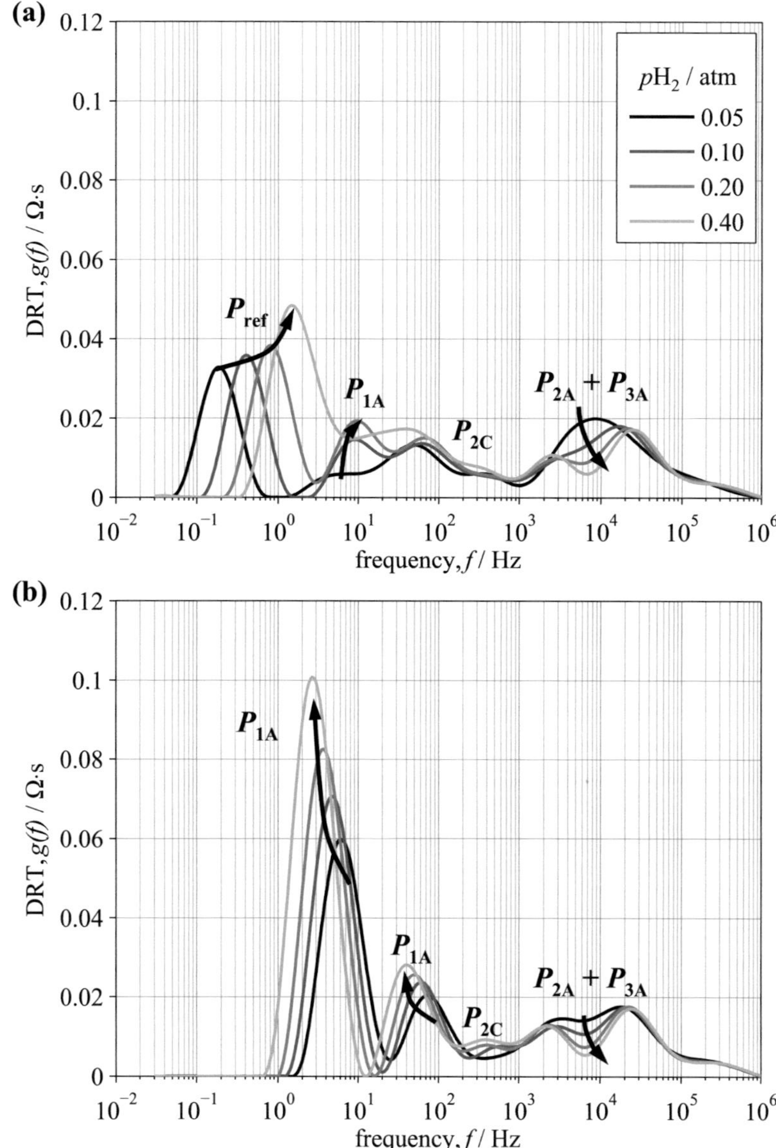

Figure 5.2.: Calculated DRT of the EIS spectra recorded for varying $pH_{2,an}$ (Fig. 5.1). (a) Reformate and (b) hydrogen operation. [cell# Z2_190]

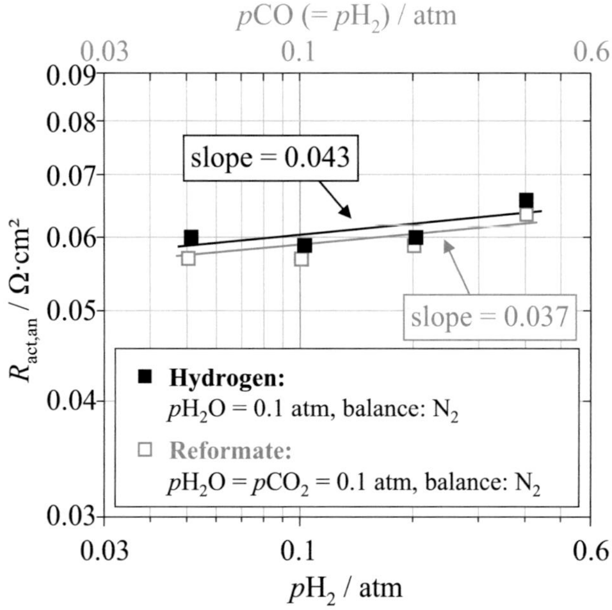

Figure 5.3.: Characteristic dependence of the anode activation polarization resistance ($R_{act,an} = R_{2A} + R_{3A}$) on the hydrogen partial pressure in the fuel gas (pH_2). [cell# Z2_190]

In the DRT plot, the peaks attributed to the anode activation polarization process (P_{2A} and P_{3A}) exhibit the same qualitative behavior for reformate and hydrogen operation. With increasing H_2 concentrations, the area under both peaks increases slightly, resulting in a stronger overlap of both peaks. Both characteristic frequencies and area under the peaks are in a similar size for hydrogen and reformate operation. The complex nonlinear least squares (CNLS) fit results deliver quantitative comparison of the anode activation polarization resistances $R_{act,an}$ (Fig. 5.3).

According to the chemical equilibrium of the applied model reformate fuel, the concentration of H_2 had to be varied along with the concentration of CO. Therefore, the values obtained for reformate operation have to be referred to $pH_{2,an}$ as well as to pCO_{an}, as accounted for in Fig. 5.3. It is evident that the area specific resistance (ASR) of the anode activation polarization is in the same range for hydrogen and reformate operation. The small deviation between both data sets (less than 3.5%) lies within the error range of the applied equivalent circuit. As expected from Fig. 5.1 and Fig. 5.2, $R_{act,an}$ shows low sensitivity on the variation of $pH_{2,an}$ in both cases. In the double-logarithmic plot, the parameter dependence can be represented by a linear trend, indicating a power-law dependence of $R_{act,an}$ on the partial pressure of H_2. The slopes of 0.043 and 0.037 correspond to the exponent in the power-law relationship between $R_{act,an}$ and $pH_{2,an}$. The relevance of this result with regards to the mechanism of the electrochemical fuel oxidation in the investigated anode will be discussed in Section 5.2.

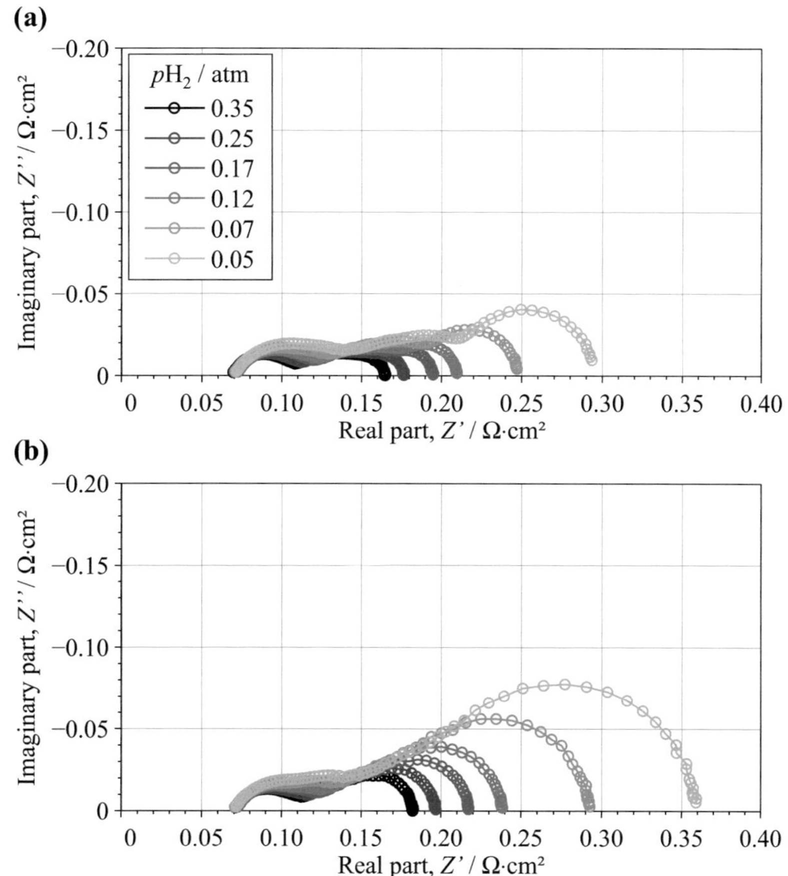

Figure 5.4.: Nyquist representation of measured EIS spectra for varying $p\mathrm{H_2O_{an}}$. (a) Reformate and (b) hydrogen operation (fuel compositions see Table 5.1; cathode: air; $T = 800\,°\mathrm{C}$). [cell# Z2_190]

5.1.2. H₂O Partial Pressure Dependence

In order to analyze the characteristic dependence of the anode activation polarization processes P_{2A} and P_{3A} on $p\mathrm{H_2O_{an}}$, the $\mathrm{H_2O}$ concentration in the fuel was varied in 6 steps from 5 to 35 %. The $\mathrm{H_2}$ concentration was kept constant at 15 %. According to the thermodynamic equilibrium of the gas mixture in the anode, the $\mathrm{CO_2}$ content was varied in the same way as $\mathrm{H_2O}$, whereas the CO content (in analogy to $\mathrm{H_2}$) was kept constant (cf. Table 5.1). Figure 5.4 depicts the spectra recorded for this parameter variation.

Comparing Figs. 5.4a and 5.4b reveals that the high-frequency arcs originating from the anode activation polarization processes are of similar size for hydrogen and reformate operation. Also the same characteristic dependence on the parameter variation can be observed for both operation types: As $\mathrm{H_2O}$ concentration increases, the arc size decreases notably, indicating a decrease in activation polarization resistance. The corresponding DRTs calculated from the measured spectra, as depicted in Fig. 5.5, confirm this finding.

In the DRT plots, it can be observed that the size and characteristic frequencies of P_{2A}

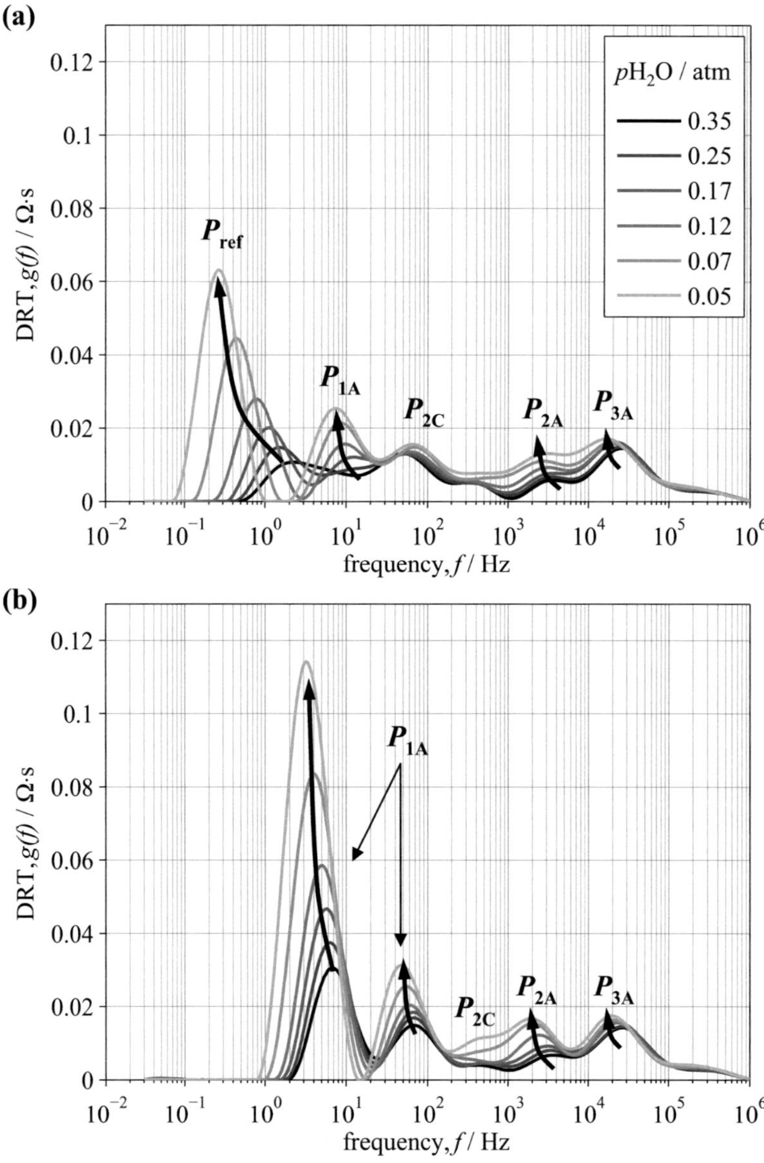

Figure 5.5.: Calculated DRT of the EIS spectra recorded for varying pH$_2$O$_{an}$ (Fig. 5.4). (a) Reformate and (b) hydrogen operation. [cell# Z2_190]

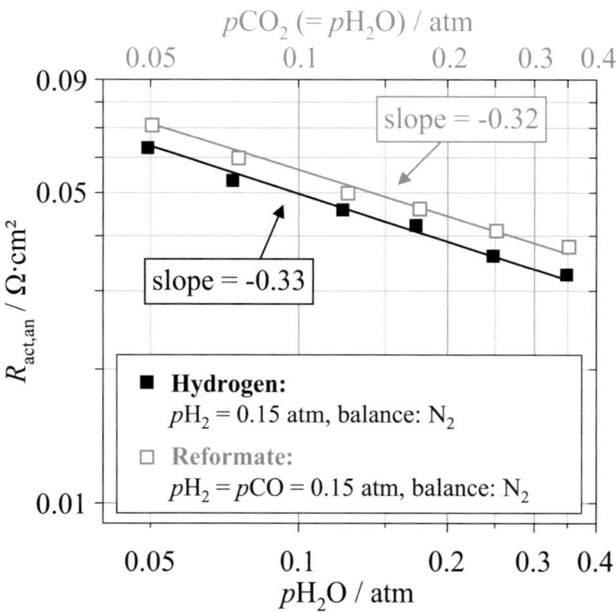

Figure 5.6.: Characteristic dependence of the anode activation polarization resistance ($R_{\mathrm{act,an}} = R_{2A} + R_{3A}$) on the steam partial pressure in the fuel gas (pH_2O). [cell# Z2_190]

and P_{3A} are in the same range for both operation types. With decreasing pH_2O_{an}, the area under both peaks increases and a shift towards lower characteristic frequencies can be observed for both operation types. The corresponding ASR of the anode activation polarization, obtained through the CNLS-fitting procedure, are depicted in Fig. 5.6.

It has to be noted that, according to the chemical equilibrium of the applied model reformate fuel, the concentration of H_2O was varied along with the concentration of CO_2. Therefore, the values obtained for reformate operation have to be referred to the partial pressure of H_2O as well as to the partial pressure of CO_2, as denoted in Fig. 5.6. For varying H_2O (and CO_2) partial pressures, the obtained anode activation polarization resistances are in the same size for both operation types. It is noted that the values measured under reformate operation are slightly (ca. 10 %) higher. Due to the content of CO and CO_2 in the reformate fuel, gas diffusion is more resistive than for hydrogen operation. Hence, the contribution of gas phase diffusion within the dense anode functional layer (AFL) to the anode activation polarization process ($P_{2A} + P_{3A}$) [87,88] explains the observed higher $R_{\mathrm{act,an}}$ for reformate operation. The observed trend for the measured ASR however is apparently not influenced by this deviation. The linear trend in the double-logarithmic plot, indicating power-law dependence of $R_{\mathrm{act,an}}$ on the partial pressure of H_2O is still clearly visible for both operation types. As for the variation of $pH_{2,\mathrm{an}}$, the slopes which correspond to the power-law exponent are virtually identical (-0.33 for hydrogen operation; -0.32 for reformate operation). The relevance of this result for the identification of the electrochemical fuel oxidation mechanism will be discussed in Section 5.2.

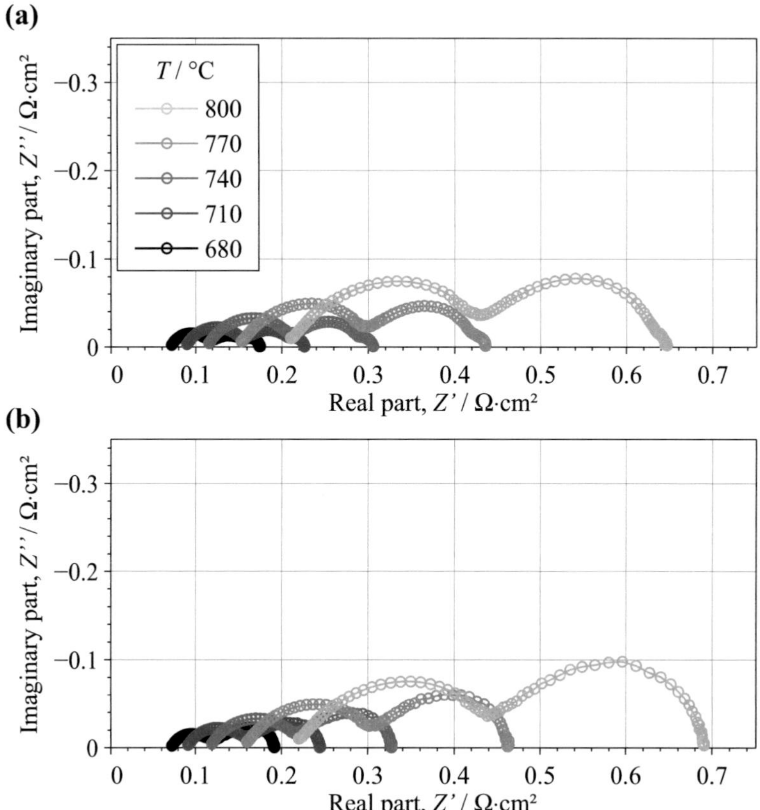

Figure 5.7.: Nyquist representation of measured EIS spectra for varying operating temperatures. (a) Reformate and (b) hydrogen operation (fuel compositions see Table 5.1; cathode: air; $T = 800\,°C$). [cell# Z2_190]

5.1.3. Temperature Dependence

In order to compare the thermal activation behavior of the electrochemical processes related to the electrochemical fuel oxidation within the anode, the cells were tested at temperatures between 680 and $800\,°C$ in steps of $30\,K$. The according equilibrium fuel gas compositions are given in Table 5.1. Figure 5.7 depicts the impedance spectra recorded for this temperature variation.

For both operation modes, thermal activation of the high-frequency arcs is clearly observable. At each temperature point, shape and size are similar for hydrogen and reformate operation.

Figure 5.8a gives the DRTs calculated from the EIS measurements operated on reformate for the five different operating temperatures. The strong temperature dependence of the processes P_{2A} and P_{2A} can be seen immediately. Along with the significant increase in resistance, the characteristic frequencies are decreasing with decreasing operating temperature. Also thermal activation of the process P_{2C} (activation polarization of the cathode) can be observed, which at lower temperatures completely covers the anodic gas diffusion process P_{1A}.

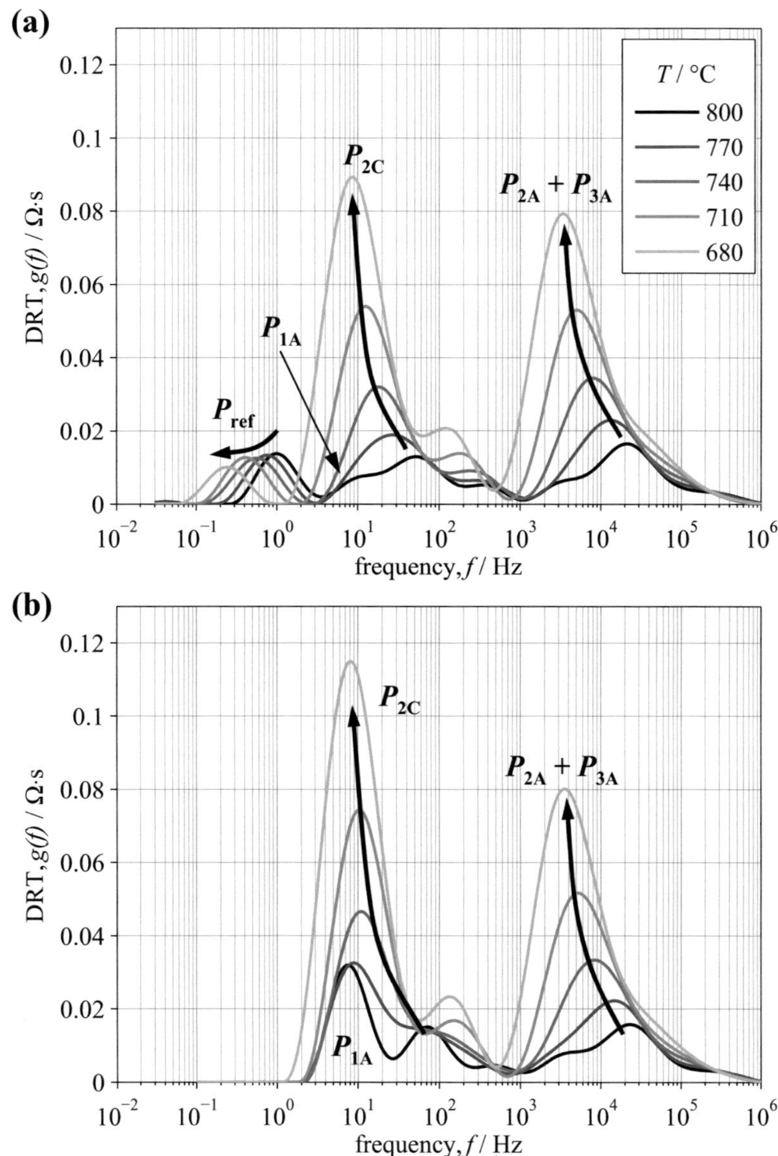

Figure 5.8.: Calculated DRT of the EIS spectra recorded for varying operating temperatures (Fig. 5.7). (a) Reformate and (b) hydrogen operation. [cell# Z2_190]

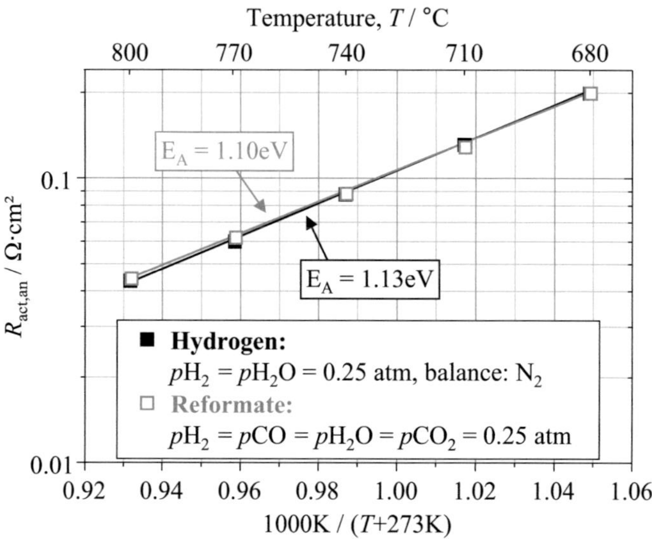

Figure 5.9.: Characteristic dependence of the anode activation polarization resistance ($R_{\mathrm{act,an}} = R_{2A} + R_{3A}$) on the operating temperature T. [cell# Z2_190]

Figure 5.8b shows the DRTs for hydrogen operation. The processes P_{2A} and P_{3A} exhibit the same temperature dependence as observed under reformate operation. Compared with Fig. 5.8a, both characteristic frequencies and size of the peaks exhibit the same characteristic trends. Again, thermal activation behavior of the cathode activation polarization (P_{2C}) can be observed here. For low temperatures, the area under the peak of P_{2C} becomes notably greater compared to reformate operation. This has to be attributed to the overlap with P_{1A} in the DRT plot. For hydrogen operation, the process P_{1A} is more pronounced than for reformate operation and hence has a greater contribution to the low-frequency arc.

The polarization resistances obtained from the CNLS-fitting are depicted in the Arrhenius plot in Fig. 5.9. As expected from the DRT (Fig. 5.8), it is now quantitatively proven that the anode activation polarization resistances are in the same size for both operation types. The activation energies, which can be deducted from these data, are in the same size for hydrogen and reformate operation ($E_{\mathrm{act,an,H_2}} = 1.13$ eV; $E_{\mathrm{act,an,ref}} = 1.10$ eV). The significance of this result with respect to the electrochemical fuel oxidation mechanism will be discussed in the following section.

5.2. Kinetic Interpretation

The coupling of the fuel oxidation kinetics and the anode activation polarization is given by the exchange current density at the anode (cf. Section 2.5.2). The kinetics of the electrochemical fuel oxidation at the anode is described by the anodic exchange current density (cf. Section 2.5.2). For EIS measurements at open circuit condition (OCC) and sufficiently small stimulus currents, as to cause a linear voltage response, the resistance of the anode activation polarization $R_{\mathrm{act,an}}$ is related to the anodic exchange current density $i_{0,\mathrm{an}}$ by the following equation [59]:

$$i_{0,\mathrm{an}} = \frac{1}{R_{\mathrm{act,an}}} \cdot \frac{RT}{2F},$$ (5.1)

where R denotes the universal gas constant, T the operating temperature and F the Faraday constant. Leonide [59] has demonstrated that for hydrogen operation, the anodic exchange current density can be described by the following semi-empiric relation [79]:

$$i_{0,\mathrm{an}} = \gamma_{\mathrm{an}} \cdot (p\mathrm{H}_{2,\mathrm{an}})^a \cdot (p\mathrm{H}_2\mathrm{O}_{\mathrm{an}})^b \cdot \exp\left(\frac{-E_{\mathrm{act,an}}}{RT}\right).$$ (5.2)

Equation 5.2 can be interpreted as a kinetic rate expression of the electrochemical oxidation of hydrogen at the anode. According to the expression, the electrochemical oxidation reaction is characterized by power-law dependencies on the partial pressures of H_2 and H_2O such as an Arrhenius type dependence on the operating temperature. Also in this study, the resistance of the anode activation polarization varied with the H_2 and H_2O partial pressures in a power-law fashion and exhibited Arrhenius type thermal activation behavior.

From the perspective of chemical kinetics, the exponent a can be interpreted as the reaction order of the electrochemical fuel oxidation with respect to H_2. The parameter b can be regarded as the order of reaction with respect to H_2O. Combining Eqs. 5.1 and 5.2 reveals that the parameters a and b are equal to the negative values of the slopes in Fig. 5.3 and Fig. 5.6, respectively [7, 59]. Furthermore, the activation energy of the anode activation polarization process obtained from the Arrhenius-plot in Fig. 5.9 can be regarded as activation energy of the anode fuel oxidation reaction. Table 5.4 lists the kinetic parameters obtained in this work in comparison to kinetic parameters reported in literature. The data from literature was obtained for the same type of anodes as applied herein [59, 60]. Similar to this work, the kinetic parameters were obtained via EIS measurements conducted under OCC.

It has to be noted that the performed variation of the H_2 concentration went along with a variation of the concentration of CO, which could in theory also be electrochemically oxidized. Thus strictly speaking, the parameter a could also be ascribed to the reaction order of the electrochemical CO oxidation with respect to CO. In analogy, the parameter b could also be ascribed to the reaction order with respect to CO_2.

However, it is evident that the kinetic parameters obtained under reformate operation are virtually identical with the kinetic parameters for hydrogen operation, while the values

Table 5.4.: Kinetic parameters for the electrochemical fuel oxidation in anode supported SOFCs.

operation type	a	b	E_{act} (eV)	reference
reformate	-0.04	0.32	1.10	this work
hydrogen	-0.04	0.33	1.13	this work
hydrogen	-0.10	0.33	1.09	[59]
carbon monoxide	-0.06	0.25	1.23	[60]

representing the electrochemical oxidation of CO are considerably different. It has hence been demonstrated unambiguously that only H_2 is oxidized electrochemically within the SOFC anode.

This finding explains the observed evolution of the anode activation polarization processes for varying H/C-ratios of the reformate fuel, as given in Fig. 4.5. Only when the cell is operated with pure CO/CO_2, a distinct increase of the anode activation polarization processes P_{2A} and P_{3A} can be observed. In contrast, the small increase with decreasing H/C-ratio of the fuel implies that even for $90\,\%$ CO/CO_2 in the fuel, the electrochemical oxidation of H_2 is still the dominating mechanism.

The observed strong preference of H_2 for the electrochemical fuel oxidation reaction implies further that the CO in the fuel must subsequently be thermally oxidized via reforming chemistry. The water-gas shift (WGS) reaction (Eq. 2.37) is reported to proceed very fast at the catalytically active Ni-sites adjacent to the gas pores within the anode [124, 126] and converts CO and H_2O into H_2 and CO_2. The interaction between reforming chemistry, in particular the WGS-reaction, and gas transport within the anode substrate will be discussed in more detail in the following chapter.

5.3. Conclusions

A detailed study by means of EIS was conducted in order to understand the mechanism of the electrochemical fuel oxidation for reformate fueled anode supported SOFCs. In earlier studies, the uncertainty of the actual fuel composition at the electrochemically active sites within the SOFC anode and missing knowledge about the polarization processes that form the measured impedance spectra inhibited detailed analyses.

The measurement procedure applied herein allows for the direct comparison of the anodic activation polarization process for reformate and hydrogen operation under defined gas compositions. The processes are compared for a systematic variation of operating parameters (H_2 partial pressure $pH_{2,\mathrm{an}}$, H_2O partial pressure pH_2O_{an} and operating temperature T). For each variation of the characteristic parameters, it is demonstrated that the impedance of the anodic activation polarization is in the same size. Consequently, the anodic activation polarization, which incorporates the electrochemical fuel oxidation, shows the same characteristic parameter dependence for both operation types.

The accuracy of the physically motivated CNLS-fit to the measured spectra led to the determination of the kinetic parameters of the electrochemical fuel oxidation at the anode. The obtained kinetic parameters are virtually the same for reformate and hydrogen operation. The results exhibit consistency with figures from literature reported for hydrogen operation [59] and show a significant difference to kinetic parameters of the electrochemical oxidation of CO [60]. Hence, it has been demonstrated unambiguously that for reformate fueled anode supported SOFCs, exclusively H_2 is electrochemically oxidized.

6. Gas Transport and Reforming Reactions

In this chapter, the gas transport properties within reformate fueled Ni/YSZ anode structures for anode supported solid oxide fuel cells (SOFCs) are analyzed via finite element method (FEM) modeling. In the measured impedance spectra, a yet unidentified coupling of gas transport and reforming chemistry manifests in form of the low-frequency polarization processes P_{1A} and P_{ref}. Based upon the findings presented in Chapters 4 and 5, a schematic understanding of the gas transport pathways within the reformate fueled anode is developed and implemented in a FEM model. As the model is able to reproduce the characteristics of the measured spectra, it is used to elucidate the physical origin of the polarization processes P_{1A} and P_{ref}. Parts of the results presented in this chapter have been elaborated in the diploma thesis of Helge Geisler [157], which was conducted under my guidance. The results presented in this chapter have partly been published in Ref. [158].

6.1. Model Development

In the previous chapter, the electrochemical oxidation mechanism of reformate fuels within Ni/YSZ anodes was investigated by a kinetic analysis. It has been demonstrated that exclusively H_2 is electrochemically oxidized at the three phase boundary (TPB), which means that CO is thermally converted via the water-gas shift (WGS) reaction (cf. Eq. 2.37) on the catalytically active Ni surface within the anode substrate. Consequently, gas transport in reformate fueled SOFC anodes features two transport pathways (H_2/H_2O and CO/CO_2) coupled by the WGS reaction, as illustrated in Fig. 6.1.

Numerical simulation can be a valuable tool in understanding the processes inside porous electrode structures, which generally exhibit coupling of various physical phenomena. In SOFC literature, numerical modeling has been broadly applied in the field of internal reforming, accurately computing the coupling of reforming chemistry, gas transport and charge transfer chemistry within Ni/YSZ anodes [15,119–126]. Also transient impedance modeling has proven to give interesting insights into the underlying coupled electrochemical reactions and species transport [128,159,160]. An implementation of the schematic model discussed above (Fig. 6.1) in numerical simulation is hence promising to contribute to the understanding of the measured low-frequency polarization processes.

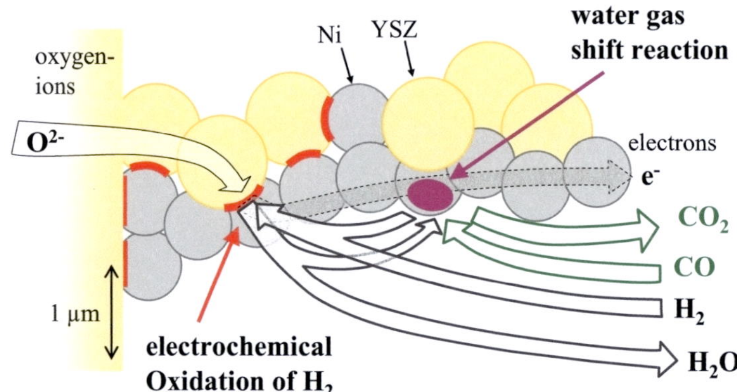

Figure 6.1.: Illustration of the reaction and transport processes for Ni/YSZ anodes in anode supported SOFCs operated with reformate fuels.

In this study, a transient FEM model is set up representing the reaction and transport processes for anode supported SOFC operating with reformate fuels as given in Fig. 6.1: (i) electrochemical oxidation of H_2, (ii) CO-conversion via the WGS reaction and (iii) gas transport via the H_2/H_2O and CO/CO_2 diffusion pathways. With the developed FEM model, it is aimed to investigate the relation between the reported characteristics of measured impedance response and the underlying gas transport and reforming processes within reformate fueled Ni/YSZ anodes.

6.2. Model Setup

The physical model represents charge-transfer chemistry, gas transport and reforming chemistry through the thickness of the anode substrate. Porous-media transport within the anode substrate is represented by the Stefan-Maxwell model [161, 162]. Heterogeneous (catalytic reforming) chemistry on the Ni-surfaces is modeled with a global reaction mechanism [120]. Charge-transfer chemistry at the electrode-electrolyte interface is modeled with a simple time-dependent rate equation. The gas compositions within the anode fuel gas channel are specified as boundary conditions of the present model. According to the experimental data, the model is set up for operation with a model reformate gas mixture consisting of H_2, H_2O, CO, CO_2 and N_2 at chemical equilibrium (cf. Chapter 3). The model was developed and implemented in the software package COMSOL [163].

Due to (i) the high flow rates of the fuel gas (250 sccm), (ii) the relatively small electrode area ($10 \times 10\,\text{mm}^2$) and (iii) the negligible fuel utilization during the impedance measurements at open circuit voltage (OCV), concentration gradients perpendicular to the electrode thickness can be neglected in the experimental setup. Therefore, the model geometry is reduced to the thickness of the anode substrate. Concentration gradients within the fuel gas channel are neglected and hence the gas compositions of the fuel supply are specified as boundary conditions at the interface between anode substrate and gas channel. The species conversion through the electrochemical charge transfer reaction is specified as boundary condition (mass flux over boundary) at the anode-electrolyte interface.

The applied mass balances account for diffusive multi-component mass transport calculated via the Stefan-Maxwell equation [161–163]:

$$\frac{\partial}{\partial t}(\rho \omega_k) + \nabla \left[-\rho \omega_k \sum_l D_{kl}^{\text{eff}} \cdot (\nabla x_l) \right] = R_{\text{el},k} + R_{\text{sh},k}. \tag{6.1}$$

In Eq. 6.1, ρ is the density of the fuel gas, ω_k is the mass fraction of the gas component k of the anode fuel, D_{kl}^{eff} is the effective binary diffusion coefficient and x_l the mole fraction of species l, with $k, l \in \{H_2, H_2O, CO, CO_2, N_2\}$. The source term includes the reaction rates of the electrochemical fuel oxidation at the anode-electrolyte interface ($R_{\text{el},k}$) and the reaction rates of the WGS reaction ($R_{\text{sh},k}$). The mass balances are calculated for $k - 1$ species. Dalton's law is applied as

$$\sum_k \omega_k = 1. \tag{6.2}$$

The effective binary diffusion coefficients incorporate the structural parameter ψ_{an}, which describes the ratio of porosity and tortuosity of the anode substrate [89, 90]. Within the pores of SOFC anodes, molecular gas diffusion is accompanied by Knudsen diffusion. Since the mass balances (Eq. 6.1) do not explicitly deal with Knudsen diffusion, an additional factor ξ_{kn}, accounting for the resistance caused by this phenomenon, is introduced:

$$D_{kl}^{\text{eff}} = \xi_{\text{kn}} \cdot \psi_{\text{an}} \cdot D_{kl}. \tag{6.3}$$

For the porous anode substrate investigated herein, a structural parameter of $\psi_{\text{an}} = 0.13$ has been determined [59]. In this work, the Knudsen-factor ξ_{kn} is fitted to experimental data. This procedure is explained in detail below. The binary diffusion coefficients D_{kl} are calculated using the Chapman-Enskog theory [89, 164].

Charge transfer chemistry at the TPB, which in fact has a penetration depth of several µm in the anode [88], is considered to take place at the anode-electrolyte interface. The reaction rate of the electrochemical fuel oxidation is represented by a simple time-dependent rate equation as

$$R_{\text{el},k} = \nu_{\text{el},k} \cdot \frac{M_k}{2F} \cdot i_0 \cos\left(2\pi f\right), \tag{6.4}$$

where $\nu_{\text{el},k}$ is the stoichiometric coefficient for species k involved in the charge transfer reaction, M_k the molar mass of species k, F is the Faraday constant, i_0 the amplitude and f the frequency of the time-dependent stimulus current. Among the species of the reformate fuel, exclusively H_2 is electrochemically oxidized (cf. Chapter 5). The resulting stoichiometric coefficients are listed in Table 6.2.

For simulation of the WGS reaction (Eq. 2.37), the following global rate equation reported by Lehnert et al. [120] was applied as

$$R_{\text{sh},k} = \nu_{\text{sh},k} \cdot k_{\text{sh}} \cdot \left(x_{CO} \cdot x_{H_2O} - \frac{1}{K_{\text{sh}}} \cdot x_{H_2} \cdot x_{CO_2} \right). \tag{6.5}$$

In Eq. 6.5, $\nu_{\text{sh},k}$ denotes the stoichiometric coefficient for species k involved in the WGS reaction (cf. Table 6.2), k_{sh} the velocity constant and K_{sh} the equilibrium constant of the

Table 6.1.: Stoichiometric coefficients of the fuel gas components with regards to the electrochemical fuel oxidation ($\nu_{\mathrm{el},k}$) and the WGS reaction ($\nu_{\mathrm{sh},k}$).

species k	H_2	H_2O	CO	CO_2	N_2
$\nu_{\mathrm{el},k}$	-1	1	0	0	0
$\nu_{\mathrm{sh},k}$	1	-1	-1	1	0

reaction. Detailed data for the equilibrium constant K_{sh} over a broad temperature range can be found in literature [107].

The presented model is simulating the transient interaction of gas transport and the WGS reaction within the reformate fueled SOFC anode. For applied transient current stimulation of the cell, the model predicts the resulting concentration profiles of each fuel gas species through the thickness of the anode substrate. In order to compute the resulting electrochemical impedance spectra, the corresponding anode concentration overpotential needs to be calculated. This is done via the modified Nernst equation [59, 77, 92] as

$$\eta_{\mathrm{conc,an}} = \frac{RT}{2F} \ln \left(\frac{x_{H_2O}^{\mathrm{TPB}} \cdot x_{H_2}^{\mathrm{GC}}}{x_{H_2O}^{\mathrm{GC}} \cdot x_{H_2}^{\mathrm{TPB}}} \right), \tag{6.6}$$

where R is the universal gas constant and T is the temperature. x_k^{TPB} and x_k^{GC} denote the molar fractions of species k at the interface anode-electrolyte and at the interface anode-gas channel, respectively. Equation 6.6 is applicable for hydrogen operation (i.e., fuel gas mixtures of H_2/H_2O) [59, 92]. As demonstrated in Chapter 5, only H_2 is electrochemically oxidized in reformate fueled Ni/8YSZ anodes, which means that H_2 and H_2O are the predominant electrochemically active species. Therefore, also for reformate operation, the calculation of the anode concentration overpotential is based on the concentrations of H_2 and H_2O at the TPB and in the gas channel (Eq. 6.6).

6.3. Impact of Porous-Media Transport

As mentioned above, the gas transport equation in the present model (Eq. 6.1) does not account for Knudsen diffusion explicitly. Instead, effective diffusion coefficients with an additional factor ξ_{kn} accounting for the overall mass transport restriction due to Knudsen diffusion are applied (see Eq. 6.3). Depending on the individual molecular mass, the impact of Knudsen diffusion on the overall diffusion properties is different for each gas species. In this work, it is aimed to find an adequate overall Knudsen factor, applicable for the reformate fuel mixtures investigated herein. For operation with reformate fuels, however, the impact of the reforming chemistry on the area specific resistance (ASR) of the predicted impedance spectra is uncertain. Therefore, ξ_{kn} is fitted to impedance data measured under operation with reference gas compositions, where CO and CO_2 of the model reformate gas composition are replaced by inert N_2. This offers the advantage that H_2 and H_2O diffuse in a reformate-like atmosphere, while the reforming reactions are not occurring.

Figure 6.2 depicts the fit of the simulated ASR for different pH_2O to experimental data

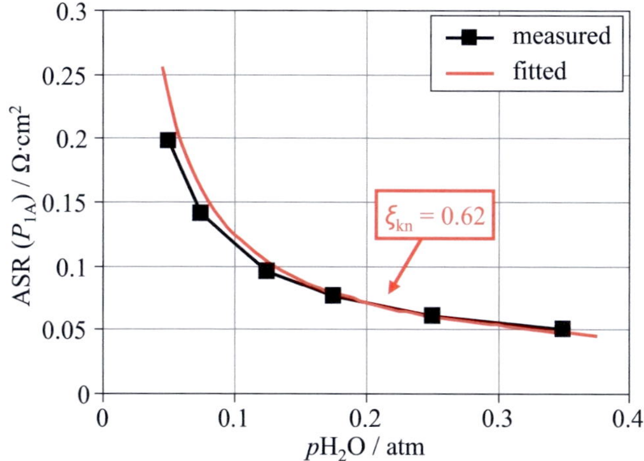

Figure 6.2.: ASR of the measured (R_{1A}) and simulated anode diffusion polarization process for varying pH$_2$O in the anode with hydrogen as fuel gas (pH$_2$ = 0.15 atm, balance N$_2$; cathode: air; T = 800 °C). [cell# Z2_190]

obtained by Electrochemical Impedance Spectroscopy (EIS) measurements.[1] The experimental ASR were obtained by complex nonlinear least squares (CNLS) fitting of the equivalent circuit model (ECM) for hydrogen operation [86]. The obtained Knudsen-factor has a value of ξ_{kn} = 0.62. The good agreement with the measured data shows that, with the obtained ξ_{kn}, the model accurately represents the resistance caused by the anode diffusion polarization process P_{1A}. Only for very low pH$_2$O (below 0.075 atm), an overestimation of the simulated ASR becomes noticeable.

Figure 6.3 illustrates a typical simulated EIS spectrum along with an EIS spectrum measured under the same operating conditions. From the comparison of the Nyquist plots (Fig. 6.3a), it is evident that the model is capable to predict Warburg-type gas diffusion behavior (cf. Section 2.5.3). The comparison of the imaginary parts of the complex impedance spectra (Fig. 6.3b) shows that the characteristic frequency of the model prediction ($f_{c,simu}$ = 19 Hz) is slightly higher than the characteristic frequency of the measured spectrum ($f_{c,meas}$ = 6 Hz).

An overall good agreement with the measurements for both stationary (ASR) and transient (EIS spectra) model predictions has been demonstrated. It has hence been shown that the model is able to accurately predict gas transport in the anode structure of the cells investigated herein.

6.4. Impact of WGS Kinetics

In the next step of this work, the impact of the WGS kinetics on the reformate diffusion impedance is investigated. The kinetics of the WGS reaction (Eq. 6.5) determines the velocity of the CO conversion, which delivers parts of the H$_2$ to be consumed in the electrochemical anode reaction (cf. Fig. 6.1). It is obvious that the WGS kinetics affects

---

[1]In the following results, the species concentrations are represented by the corresponding partial pressures in the gas channel, e.g., pH$_2$O = $x^{GC}_{H_2O} \cdot p_0$, with p_0 = 1 atm.

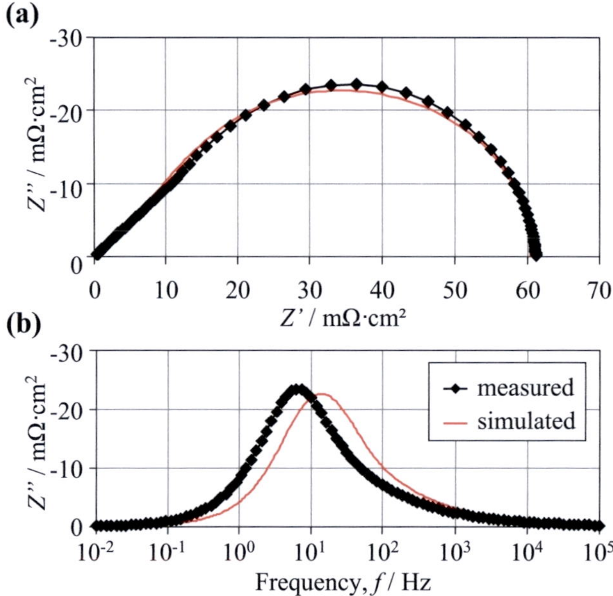

Figure 6.3.: Measured ($R_{1A} + R_{ref}$) and simulated anode diffusion polarization processes occurring under reformate operation ($pH_2 = 0.15$ atm, $pH_2O = 0.25$ atm, balance N_2; $T = 800\,°C$). (a) Nyquist representation. (b) Imaginary components of the complex impedances as functions of frequency. [cell# Z2_190]

the concentration profiles of the electrochemically active species H_2 and H_2O and thus affects the shape and the ASR of the anodic gas transport impedance.

Within SOFC literature, there is a great deal of knowledge about the kinetics of steam reforming [119–126], whereas comparatively little is known about the kinetics of the WGS reaction in SOFC anodes. Due to the much higher reaction rates compared with the steam reforming reaction, most of the reported investigations assume the WGS reaction to be in equilibrium. However, from detailed studies [120, 125, 126], it can be deducted that the velocity constant for the WGS reaction is generally one to two orders of magnitude higher than those of the reforming reaction. In this study, the WGS kinetics is represented by the global reaction rate reported by Lehnert et al. [120] (Eq. 6.5). The reversible rate expression assumes a first order reaction with respect to all the involved reactants (i.e., H_2, H_2O, CO and CO_2). The sensitivity of the simulated impedance spectra on the kinetics of the WGS reaction is illustrated in Fig. 6.4, where predicted spectra for various velocity constants k_{sh} are given.

The impact of k_{sh} on ASR and shape of the predicted impedance spectra is evident. When the velocity constant k_{sh} is set to zero, no CO is converted and consequently H_2 and H_2O diffuse in inert CO-CO_2. The resulting concentration gradients of H_2 and H_2O through the anode substrate are high, and so is the ASR of the predicted spectrum. The shape of the predicted impedance curve represents Warburg-type gas diffusion impedance (cf. Section 2.5.3). For very fast WGS kinetics ($k_{sh} = \infty$), CO and H_2O react instantaneously forming H_2 and CO_2. The resulting gradients of H_2 and H_2O are much smaller, resulting in a lower ASR. Also here, the predicted impedance spectrum exhibits Warburg-type shape.

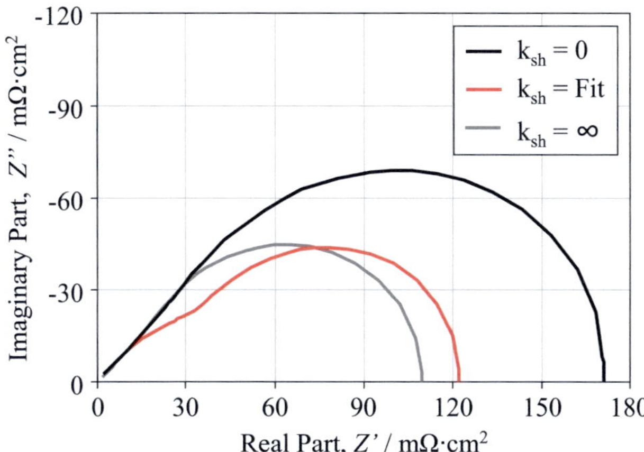

Figure 6.4.: Nyquist representation of predicted EIS spectra under reformate operation (pH$_2$ = 0.15 atm, pH$_2$O = 0.07 atm, pCO = 0.15 atm, pCO$_2$ = 0.07 atm, balance N$_2$; T = 800 °C) for varied velocity constants of the WGS reaction (k_{sh}).

For the chosen finite velocity constant of k_{sh} = 1.1×10^{-6} mol m^{-3}Pa^{-2}s^{-1} however, the shape of the predicted impedance spectrum is notably different. In accordance with the measured spectra, the EIS spectrum predicted by the transient simulation exhibits two clearly distinguishable time constants. It is noted that the chosen velocity constant is one order of magnitude higher than the value reported by Lehnert [120]. Given the high uncertainty of the kinetic data for the WGS reaction on SOFC anodes in literature, this value is still in a reasonable region. A comparison of the measured ASR with the ASR predicted for k_{sh} = 1.1×10^{-6} mol m^{-3}Pa^{-2}s^{-1} over a broad variation of pH$_2$O (and pCO$_2$) in the fuel gas is given in Fig. 6.5.

The experimental ASRs ($R_{1A} + R_{ref}$) were obtained from the CNLS-fit of the equivalent circuit as presented in Section 4.5. As it is the case for the operation with H$_2$, H$_2$O and N$_2$ (see Fig. 6.2), the model tends to predict higher ASR than the measurements at low pH$_2$O. For pH$_2$O above 0.2 atm, the simulation shows excellent agreement with the measurements. The overall good agreement between measurements and model predictions confirms the validity of the chosen WGS kinetic parameters.

Figure 6.6 illustrates a typical simulated EIS spectrum along with an EIS spectrum measured under the same operating conditions. Comparing the Nyquist plots of simulated and measured spectra (Fig. 6.6a), it is evident that both measurement and simulation exhibit the same qualitative behaviors. It has to be stated that the impedance arc at lower frequencies is more pronounced in the model prediction. It is also noticeable that the two arcs of the simulated spectra have a stronger overlap. However, it is clearly observable that the model is able to reproduce the multi-arc impedance spectra reported in literature [114, 128, 155].

Comparison of the imaginary parts of the complex impedance spectra (Fig. 6.6b) confirms the ability of the model to reproduce spectra with two different characteristic time constants. However, it is evident that the characteristic frequencies of the model predictions ($f_{c,P_{ref}}$ = ca. 5 Hz and $f_{c,P_{1A}}$ = ca. 100 Hz) are around one order of magnitude higher than

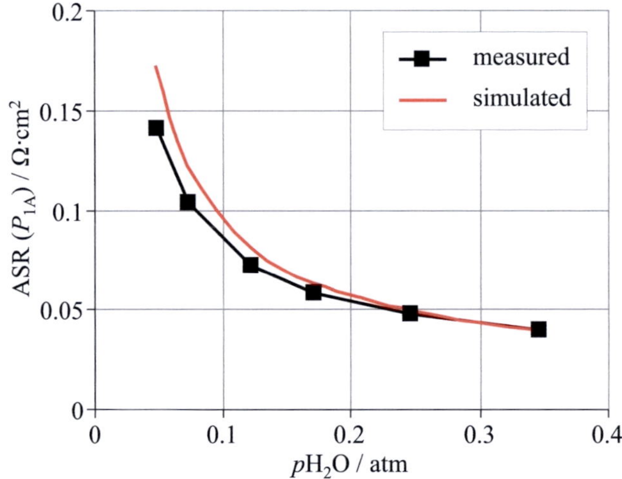

Figure 6.5.: ASR of the measured ($R_{1A} + R_{ref}$) and simulated anode diffusion polarization process for varying pH$_2$O in the anode with reformate as fuel gas (pH$_2$ = 0.15 atm, pH$_2$O = 0.07 ... 0.35 atm, pCO = 0.15 atm, pCO$_2$ = 0.07 ... 0.35 atm, balance N$_2$; T = 800 °C). [cell# Z2_190]

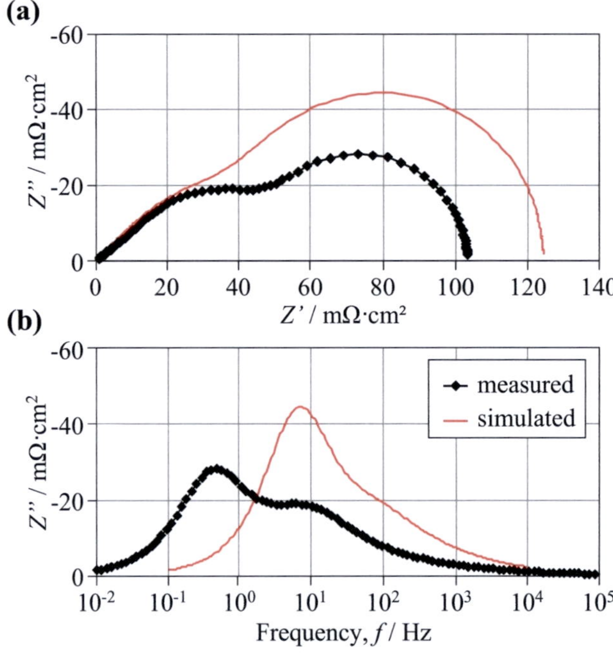

Figure 6.6.: Measured ($R_{1A} + R_{ref}$) and simulated anode diffusion polarization processes occurring under reformate operation (pH$_2$ = 0.15 atm, pH$_2$O = 0.07 atm, pCO = 0.15 atm, pCO$_2$ = 0.07 atm, balance N$_2$; T = 800 °C). (a) Nyquist representation. (b) Imaginary components of the complex impedances as functions of frequency. [cell# Z2_190]

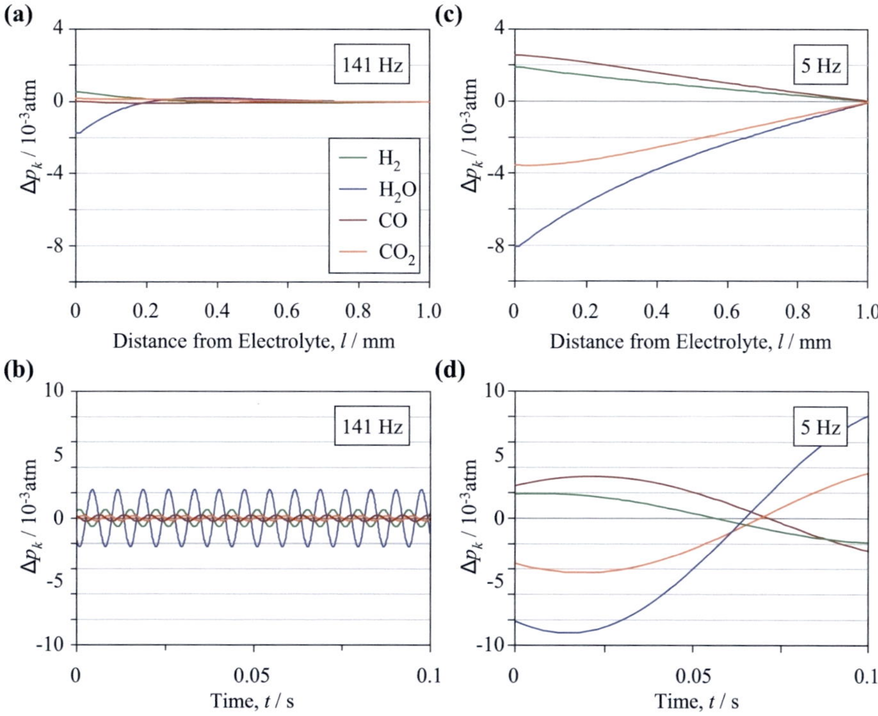

Figure 6.7.: Predicted concentration profiles and corresponding temporal evolution of the concentration gradients through the thickness of the anode substrate for excitation at (a,b) 141 Hz and (c,d) 5 Hz.

the characteristic frequencies of spectra measured under the same operating conditions.

6.5. Identification of P_{1A} and P_{ref}

For further investigation of physical origin of the resulting polarization processes ($P_{1A} + P_{ref}$), the underlying gas transport properties are analyzed with the help of the model. Figure 6.7 illustrates the predicted concentration profiles of the fuel gas species within the anode substrate. The simulation results are given for excitation with transient stimulus currents at the characteristic frequencies of the predicted polarization processes (141 Hz and 5 Hz).

For excitation at 141 Hz, the simulated concentrations of H_2 and H_2O exhibit notable concentration gradients within the first 200 μm of the anode substrate (Fig. 6.7a), indicating transport of H_2/H_2O in this region. In comparison, the concentrations of CO and CO_2 are almost constant through the whole thickness of the anode substrate. Figure 6.7b gives the evolution with time of the total concentration gradients through the electrode thickness, which confirms that at the characteristic frequency of P_{1A}, only the concentrations of H_2 and H_2O are excited by the stimulus current. Apparently the CO-conversion via the WGS reaction does not proceed in this frequency region and hence anodic gas transport is dominated by the diffusion of H_2 and H_2O.

For excitation at the characteristic frequency of the predicted second polarization process (P_{ref}, $f_c = 5$ mHz), almost linear gradients of H_2 and H_2O throughout the entire substrate are predicted (Fig. 6.7c). Interestingly, also CO and CO_2 exhibit pronounced concentration gradients through the entire thickness of the anode substrate, which however are less pronounced within the first several µm of the anode substrate. Hence, at low frequencies, the WGS reaction is proceeding and additionally to the H_2/H_2O transport pathway, reactants for the electrochemical anode reaction are now also delivered via the CO/CO_2 transport pathway. The evolution with time of the concentration gradients confirm the clearly visible transport of CO and CO_2 in this frequency region (Fig. 6.7d). Due to the fact that diffusion of CO/CO_2 proceeds much slower than H_2/H_2O diffusion, it can be concluded that the anodic gas transport impedance is dominated by the transport of CO/CO_2 in this frequency region.

It is hence demonstrated that gas transport properties within reformate fueled Ni/YSZ for anode supported SOFCs originate the measured low-frequency impedance response with two characteristic time scales. At the characteristic frequencies of the polarization process P_{1A}, the impedance response is dominated by the transport of H_2 and H_2O to and from the electrochemically active sites of the Ni/YSZ anode. At the lower frequencies of P_{ref}, CO conversion via the WGS reaction is excited and the resulting slow transport of CO/CO_2 dominates the spectrum.

6.6. Conclusions

An approach to the understanding of the gas transport properties within reformate fueled Ni/YSZ anodes in anode supported SOFCs via electrochemical impedance modeling is presented. Aim of this study was to understand the physical origin of the two overlapping low-frequency polarization processes (P_{1A} and P_{ref}), which have been identified and attributed to anode side gas transport and reforming chemistry in Chapter 4.

Based on the findings presented in Chapter 5, a schematic model of the gas transport properties in reformate fueled Ni/YSZ anode structures has been developed: Gas transport to and from the electrochemically active sites proceeds via two diffusion pathways (H_2/H_2O and CO/CO_2), which are coupled by the water-gas shift (WGS) reaction. This schematic model was implemented in a transient FEM simulation capable to coherently calculate the complex coupling of species transport phenomena and reforming kinetics. Output of the model is a transient, space-resolved prediction of the gas composition within the anode, from which EIS spectra can be simulated.

After demonstrating the ability to accurately reproduce the characteristics of the measured P_{1A} and P_{ref}, the model was used to assign the underlying physical phenomena to the polarization processes. It is demonstrated that the polarization process P_{1A} is dominated by transport of H_2/H_2O, whereas at the low characteristic frequencies of P_{ref}, the WGS reaction converts CO and facilitates additional gas transport via the CO/CO_2 diffusion pathway. Along with the results presented in Chapter 5, an overall understanding of the electrochemical polarization processes for reformate fueled anode supported SOFCs with Ni/YSZ anodes has been achieved.

7. Sulfur Poisoning

In this chapter, the impact of sulfur poisoning on the electrochemistry of reformate fueled anode supported solid oxide fuel cells (SOFCs) has been analyzed via Electrochemical Impedance Spectroscopy (EIS). Sulfur containing fuel impurities, which act as catalyst poison, are seriously affecting the performance of SOFCs. A severe degradation of the measured anodic polarization processes is reported for the operation on sulfur containing reformate. Based on the findings in the preceding chapters, this degradation can be assigned unambiguously to the poisoning of (i) the electrochemical fuel oxidation reaction at the anode functional layer (AFL) and (ii) the reforming reactions within the anode substrate. Parts of the results presented in this chapter have been elaborated in the bachelor thesis of Sebastian Dierickx [165], which was conducted under my guidance. The results presented in this chapter have been published in Ref. [166].

7.1. Operation on H_2S-free Reformate

In this study, the cell is operated with a model reformate gas containing H_2, H_2O, CO, CO_2 and N_2 at chemical equilibrium (cf. Chapter 3). After proving stability of the cell under reformate operation, contents of H_2S (0.1 and 0.5 ppm, cf. Sections 7.2 and 7.3) are added to the fuel. The degradation of the cell is monitored via EIS measurements, which are taken at regular intervals. The distribution of relaxation times (DRT) method combined with the subsequent complex nonlinear least squares (CNLS) fitting of the equivalent circuit model (ECM) presented in Chapter 4 enables a separated quantitative analysis of the single electrochemical polarization processes.

Table 7.1.: Anodic partial pressures of H_2, H_2O, CO, CO_2 and N_2 selected for the model reformate applied as anode fuel in this study ($T = 800\,°C$).

$p\mathrm{H}_{2,\mathrm{an}}$ (atm)	$p\mathrm{H}_2\mathrm{O}_{\mathrm{an}}$ (atm)	$p\mathrm{CO}_{\mathrm{an}}$ (atm)	$p\mathrm{CO}_{2,\mathrm{an}}$ (atm)	$p\mathrm{N}_{2,\mathrm{an}}$ (atm)
0.15	0.12	0.15	0.13	0.45

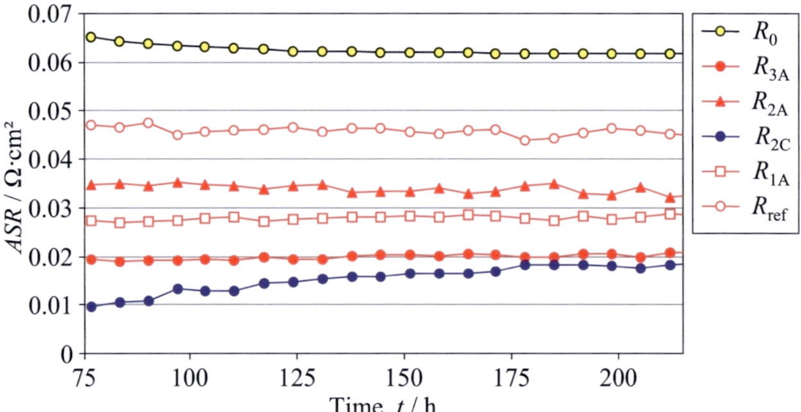

Figure 7.1.: Degradation of the ohmic resistance (R_0) and the single polarization resistances (R_{ref}, R_{1A}, R_{2C}, R_{2A} and R_{3A}) obtained by CNLS-fitting of the impedance spectra measured under H_2S-free reformate operation. [cell# Z6_127]

After the start-up procedure and initial cell testing, the cell was operated with the (H_2S-free) reformate gas mixture as given in Table 7.1 for more than 150 h. Figure 7.1 depicts the temporal evolution of the individual polarization resistances and of the ohmic resistance for this period.

During the operation with H_2S-free model reformate, no significant degradation of the polarization resistance is recorded. The relatively small decrease of the ohmic resistance ($\Delta R_0 = -3\,\text{m}\Omega\text{cm}^2$) can be explained with a slight improvement of contacting during the first 150 h of operation. The polarization process P_{2C}, which is originated by the activation polarization of the lanthanum strontium cobalt ferrite (LSCF) cathode shows a stronger degradation than the other polarization processes. This behavior is consistent with degradation analyses on the same type of cells as applied herein. The works of Endler-Schuck et al. report a strong degradation of P_{2C} during the first 200 to 300 h of operation [50, 57, 58]. The resistances of the anodic polarization processes P_{ref}, P_{1A}, P_{2A} and P_{3A} do not exhibit a notable degradation within the initial period (more than 100 h) under H_2S-free reformate operation. It can hence be concluded that the cell is stable for operation on the applied reformate fuel.

7.2. Operation on Reformate with 0.1 ppm H_2S

After proving stability of the cells under operation with the applied model reformate, contents of H_2S (0.1 ppm) were added to the fuel. Figure 7.2 displays temporal evolution of the measured ohmic and polarization losses of the cell.

It can be observed clearly that the anodic polarization processes P_{ref}, P_{1A}, P_{2A} and P_{3A}, as well as the ohmic resistance of the cell R_0, do not exhibit degradation after introduction of 0.1 ppm H_2S to the fuel. The corresponding area specific resistances (ASRs) are stable within the observed 100 h of operation. Compared to the first 200 h of operation (cf. Fig. 7.1), also the cathode activation polarization process P_{2C} shows almost stable behavior.

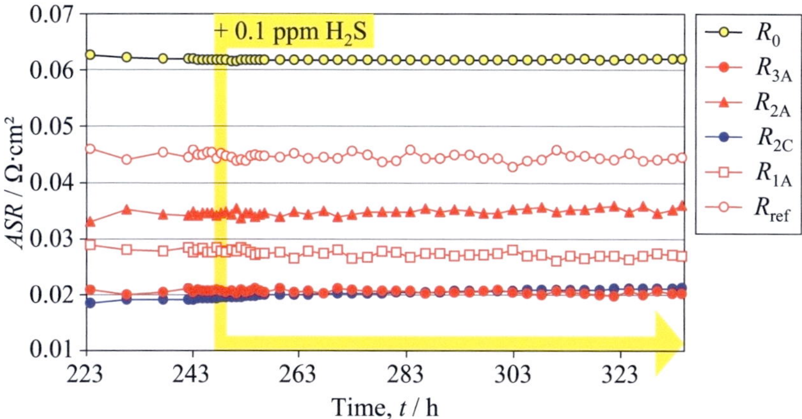

Figure 7.2.: Degradation of the ohmic resistance (R_0) and the single polarization resistances (R_{ref}, R_{1A}, R_{2C}, R_{2A} and R_{3A}) obtained by CNLS-fitting of the impedance spectra measured under reformate operation with 0.1 ppm H_2S. [cell# Z6_127]

7.3. Operation on Reformate with 0.5 ppm H_2S

In order to achieve significant cell degradation due to sulfur poisoning in this study, the H_2S concentration was increased to 0.5 ppm after the initial poisoning period of 100 h at 0.1 ppm. Immediately after the increase of the H_2S concentration to 0.5 ppm, the polarization resistance R_{pol} of the cell began to increase by $115\,\%$, as depicted in Fig. 7.3a, which gives the recorded EIS spectra for the first 17 hours after switching to 0.5 ppm H_2S. It is interesting to see that the high-frequency intersect of the impedance spectra with the real axis is constant. This indicates that the ohmic resistance of the cell is not affected by the presence of sulfur.

When looking at the DRT plot (Fig. 7.3b), the observed increase in R_{pol} can be assigned to the strong increase of the polarization processes P_{1A}, P_{2A} and P_{3A}. It is also interesting to see that the low-frequency process P_{ref} is decreasing dramatically at the same time. The peak of the polarization process P_{2A}, which is linked to the electrochemical fuel-oxidation, shows the strongest increase. Further it can be seen that, except for P_{1A}, all the anodic polarization processes shift to lower characteristic frequencies, indicating a slow-down of the underlying physical processes.

Figure 7.4 gives the ASRs of the single polarization mechanisms obtained by the CNLS-fitting. The degradation of the anodic polarization mechanisms exhibits a dramatic increase within the first 24 h of the exposure to 0.5 ppm H_2S. After 24 h, the drastic initial increase of the anodic polarization processes levels off and the corresponding losses seem to have reached constant values. Table 7.3 summarizes the resulting ASRs of the individual loss contributions.

As expected, the cathode activation polarization resistance (R_{2C}) has not been affected by the sulfur poisoning. The CNLS-fit results confirm further that also the ohmic resistance of the cell (R_0) is constant. It is noted that the decrease of the cathode activation resistance during the poisoning is most probably due to errors of the fit. However, compared to the amount of the overall polarization resistance, the obtained decrease of R_{2C}

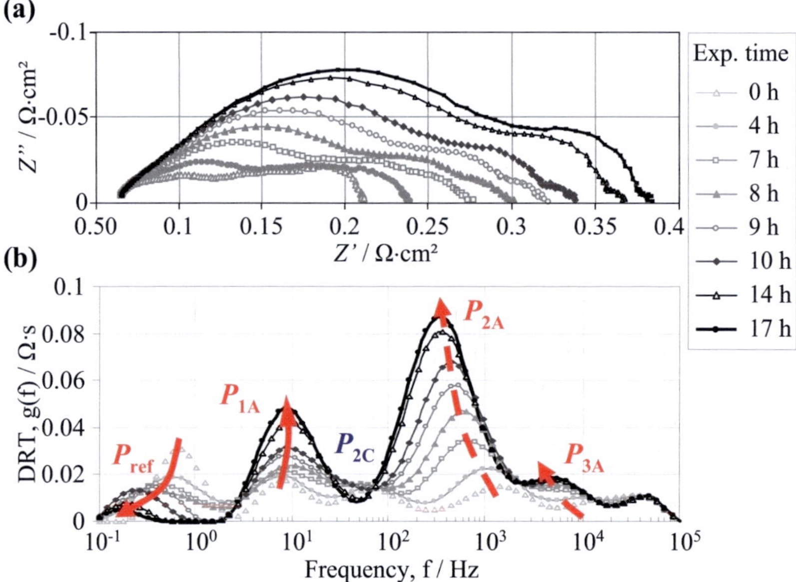

Figure 7.3.: Series of impedance spectra recorded under operation with $0.5\,\mathrm{ppm}$ H_2S in the reformate fuel: (a) Nyquist representation and (b) corresponding DRT. [cell# Z6_127]

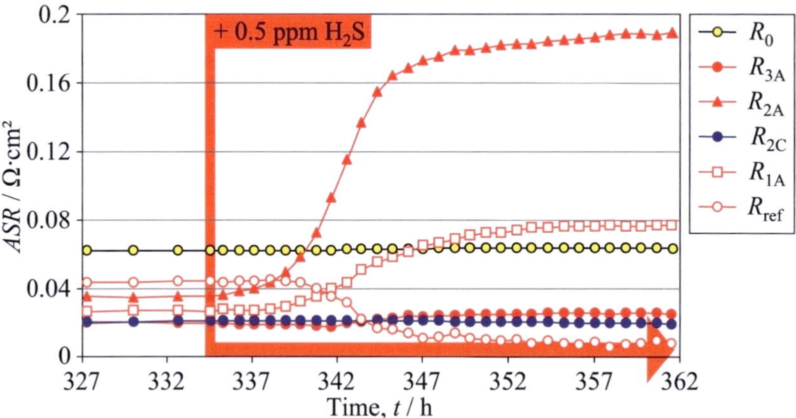

Figure 7.4.: Degradation behavior of the ohmic resistance (R_0) and the single polarization resistances (R_{ref}, R_{1A}, R_{2C}, R_{2A} and R_{3A}) obtained by CNLS-fitting of the impedance spectra measured under reformate operation with $0.5\,\mathrm{ppm}$ H_2S. [cell# Z6_127]

Table 7.2.: Area-specific resistances of the individual loss contributions before ($t = 335\,\mathrm{h}$) and after ($t = 362\,\mathrm{h}$) introduction of 0.5 ppm H_2S to the reformate fuel.

loss contribution	ASR ($t = 335\,\mathrm{h}$) ($\mathrm{m\Omega cm^2}$)	ASR ($t = 362\,\mathrm{h}$) ($\mathrm{m\Omega cm^2}$)	increase (%)
R_0	62.1	63.7	2.6
R_{2A}	35.9	189.7	428.4
R_{3A}	20.2	25.5	26.7
$R_{\mathrm{act,an}}$ ($R_{2A} + R_{3A}$)	56.1	215.3	284.0
$R_{\mathrm{act,cat}}$ (R_{2C})	21.2	19.5	-7.8
R_{1A}	27.0	77.4	186.5
R_{ref}	44.5	7.8	-82.5
$R_{\mathrm{diff,an}}$ ($R_{1A} + R_{\mathrm{ref}}$)	71.5	85.1	19.1
R_{pol}	148.7	319.9	115.2

by $1.66\,\mathrm{m\Omega cm^2}$ can be neglected.

The anode activation polarization process P_{2A} exhibits the strongest degradation during the investigated time span. The value of the corresponding ASR increases by more than five times from 35.9 to $189.7\,\mathrm{m\Omega cm^2}$. Also the polarization process P_{3A} exhibits notable degradation (R_{3A} increases by 27 %). For the anode activation polarization resistance ($R_{\mathrm{act,an}} = R_{2A} + R_{3A}$), this means an increase by almost factor four.

Also the trends observed for the low-frequency processes of the measured spectra, are represented clearly by the CNLS-fit results. The polarization process P_{1A} increases by almost factor three (R_{1A} increases from 27.0 to $77.4\,\mathrm{m\Omega cm^2}$), whereas the polarization process P_{ref} almost vanishes (R_{ref} decreases from 44.5 to $7.8\,\mathrm{m\Omega cm^2}$). With R_{1A} increasing and R_{ref} decreasing, the degradation of the low-frequency processes has a minor contribution to the observed overall increase of R_{pol}. Compared to $R_{\mathrm{act,an}}$, the sum of both low-frequency resistances (referred to as $R_{\mathrm{diff,an}}$ in Table 7.3) is almost constant (increase by 19 %) during the observed poisoning period.

7.4. Observed Poisoning Mechanisms

In the following sections, the observed impact of sulfur poisoning on the high-frequency and the low-frequency anode polarization processes will be discussed in detail separately. Based on the findings presented in the preceding chapters, it will be shown that the present results enable an assignment to the poisoning of (i) electrochemical and (ii) reforming reactions. It has to be noted that, since the EIS measurements under operation with H_2S containing fuels were taken at open circuit voltage (OCV), the poisoning effects reported herein are very high. In real operation, when the SOFC is exposed to high current densities, the poisoning effects are expected to be less pronounced (cf. Section 2.6.3). Notwithstanding that, it has to be pointed out that the qualitative trends observed herein are representative for sulfur poisoning of Ni/YSZ anodes for anode supported SOFCs.

Unlike reported in literature (cf. Section 2.6.3), no degradation was observed for the ohmic resistance of the cell R_0 during the H_2S poisoning in this study (cf. Table 7.3).

This implies that the sulfur does not affect the electronic and ionic conductivities of the investigated cell. Hence, the dominating poisoning mechanism must be ascribed to the blocking of catalytically active sites on the Ni-surface within the anode by chemisorption of sulfur (cf. Section 2.6.3).

A method to estimate the impact of the sulfur concentration on the electrochemistry of SOFC-anodes is provided by Hansen [167]. He found that the chemisorbed sulfur on the Ni surface did not show any effect on the electrochemical performance for sulfur coverages smaller than approx. 0.6. As given in Ref. [167], the sulfur coverage θ_s for chemisorbed sulfur on Ni can be calculated by the following equation:

$$\theta_s = 1.45 - 9.53 \cdot T + 4.17 \cdot 10^{-5} \cdot T \cdot \ln\left(\frac{p\mathrm{H_2S_{an}}}{p\mathrm{H_{2,an}}}\right) \tag{7.1}$$

The equation is derived from a Temkin-like adsorption isotherm proposed by Alstrup et al. [168], which has been used to correlate data on sulfur coverage on Ni surfaces. For the applied fuel gas composition with a H_2S concentration of 0.1 ppm, the sulfur coverage of the Ni surfaces within the anode structure reaches a value of $\theta_s = 0.71$. Hence, it must be concluded that the observed stability in reformate with 0.1 ppm H_2S for a period of 100 h does not guarantee a similar stability for longer periods of time. The applied cell is an anode supported cell with a substrate thickness of 1 mm. The cell hence provides a relatively high Ni surface area disposable for sulfur coverage. This might explain the fact that the applied H_2S concentration did not lead to a notable degradation of the anodic polarization processes during the observed time (100 h). Based on this experiment it should not be stated that the cell is stable at a sulfur concentration of 0.1 ppm. For a detailed analysis of the impact of the applied H_2S-concentration (0.1 ppm) on the degradation of the cell, a long-term study is necessary.

For the exposure to 0.5 ppm H_2S, the sulfur coverage reaches a value of $\theta_s = 0.78$ according to Eq. 7.1. As the drastic initial increase of the anodic polarization processes levels off after 24 h, it can be assumed that the sulfur coverage of the Ni within the anode has reached steady state during this time span.

7.4.1. Poisoning of the Electrochemical Fuel Oxidation

The strongest degradation under operation with 0.5 ppm H_2S is experienced by the anode activation polarization resistance $R_{\mathrm{act,an}} = R_{2A} + R_{3A}$. The corresponding polarization process is coupled to the electrochemical fuel oxidation within the AFL (cf. Chapter 5). The observed degradation is attributed to the chemisorption of sulfur on the catalytically active Ni sites [135, 167], which results in a reduced number of catalytically active sites and hence a retarded reaction rate of the electrochemical fuel oxidation [129, 137, 139, 167, 169]. The retarded reaction rate leads to the observed increase in resistance for the polarization processes P_{2A} and P_{3A} along with decreasing corresponding characteristic frequencies during the H_2S-exposure (Fig. 7.3b).

Detailed analyses have shown that the anodic activation polarization processes P_{2A} and P_{3A} are formed by the complex coupling of ionic transport, gas transport and the electrochemical fuel oxidation [7, 86–88]. The occurrence of these coupled mechanisms has a certain spatial extension starting from the interface of electrolyte and AFL towards the anode substrate. For the anodes applied in this study, it has been demonstrated that this penetration depth lies within the dimensions of the AFL, which has a thickness of 7 μm [7,88].

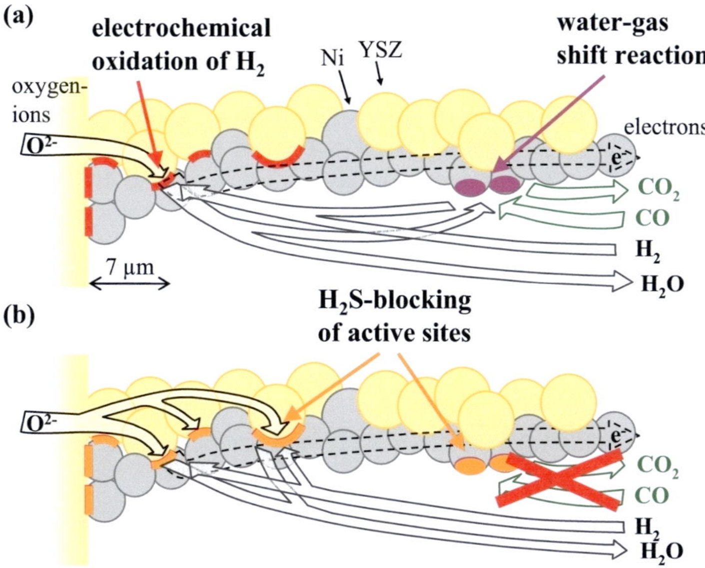

Figure 7.5.: Illustration of the reaction and transport processes in reformate fueled Ni/YSZ anodes: (a) H_2S free operation (unpoisoned) and (b) operation with 0.5 ppm H_2S.

However in the case of a dramatically lowered electrochemical reaction rate due to sulfur poisoning, a considerable extension of the penetration depth of the electrochemical processes has to be assumed (Fig. 7.5). In particular, the concomitant extension of the ionic transport path explains the observed dramatic increase in $R_{act,an}$.

From these findings, several methods can be deduced in order to reduce the effect of sulfur poisoning of the electrochemical fuel oxidation in Ni/8YSZ anodes:

- The application of highly ionic conducting materials in the electrolyte phase of the anode: The discussed extension of the ionic transport path within the anode would have a lower impact on the overall increase in R_{pol}. This explains the fact that nickel-based anodes with scandia stabilized zirconia (SSZ), which exhibits a higher ionic conductivity than yttria stabilized zirconia (YSZ), are more tolerant to sulfur poisoning [137, 142].

- The development of anode microstructures with a higher three phase boundary (TPB) density: A higher amount of electrochemically active sites within the AFL would diminish the effect of the increasing penetration depth.

- The incorporation of a sulfur tolerant catalyst at the electrochemically active sites at the anode. Recent development of metal supported SOFCs has demonstrated that it is possible to impregnate SOFC anodes with electrocatalysts such as CeO_2 [45, 46].

7.4.2. Poisoning of the Reforming Reactions

In Chapter 5, it has been demonstrated that for reformate fueled anode supported cells (ASCs), only H_2 is electrochemically oxidized at the AFL, whereas the CO within the fuel is subsequently oxidized via the water-gas shift (WGS) reaction. Accordingly, the resulting gas transport properties within the anode substrate feature two transport pathways (H_2/H_2O and CO/CO_2), which are coupled by the catalytically activated WGS reaction. The impedance response of this coupled system is given by the two low-frequency arcs, which are treated as the polarization processes P_{1A} and P_{ref}. In Chapter 6, it has been demonstrated that P_{1A} is assigned to the H_2/H_2O diffusion pathway and P_{ref} is assigned to the CO/CO_2 diffusion pathway within the anode substrate. During the exposure to 0.5 ppm H_2S, it is clearly observable that the impedance arc P_{1A} rises, whereas P_{ref} decreases (cf. Fig 7.3). The latter means that less fuel is transported via the CO/CO_2 diffusion pathway. This can only be explained by a decreased CO conversion via the WGS reaction, as illustrated in Fig. 7.5. With the sulfur poisoned shift reaction, the fuel is rather diluted with the almost inert CO and CO_2, which results in an increased H_2/H_2O diffusion polarization resistance (R_{1A}).

Sulfur poisoning of the WGS reaction in Ni cermet anodes has been reported in literature [143]. However, this is the first time that this effect can be monitored by EIS measurements. It is noted that the effects of the sulfur poisoned WGS reaction represent only a minor contribution to the overall increase in the ASR of the cell (cf. Table 7.3). This result is still of great importance to the operation of the cell. For the reformate fueled SOFC examined herein, the sulfur poisoning of the shift reaction means that the cell is incapable to utilize most parts of the CO. If the cell is fueled with e.g. a diesel-reformate, the sulfur poisoning of the Ni based anode would have the consequence that almost 50% of the fuel cannot be utilized.

In order to guarantee a high fuel utilization of the cell, also for operation with H_2S containing gases, it seems most practical to apply a sulfur tolerant shift catalyst. Application in, e.g., the anode gas compartment would be sufficient to guarantee CO conversion via the WGS reaction. A more sophisticated approach would be the infiltration of sulfur tolerant catalyst materials (such as CeO_2) within the anode substrate. Along with the guaranteed CO conversion, this method would maintain the CO/CO_2 gas diffusion pathway within the anode substrate. The latter would have positive effects on the anode diffusion polarization and hence on the performance of the cell.

7.5. Conclusions

The impact of sulfur poisoning on the electrochemistry of anode supported SOFCs has been analyzed via EIS. The tested cells have been operated with a model reformate gas mixture at chemical equilibrium. During operation, contents of H_2S (0.1 and 0.5 ppm) have been added to the fuel. A detailed analysis of impedance spectra via the DRT method and the subsequent CNLS-fit of the physically motivated equivalent circuit developed in Chapter 4 enabled a separated treatment of the resulting degradation of the individual polarization processes.

The exposure to 0.1 ppm H_2S did not affect the electrochemistry of the cells in this study. No significant degradation of the polarization processes was observed within 100 h. During the exposure to 0.5 ppm H_2S however, a considerable increase of the polarization resistance of the cell by 115% was observed. It has been shown that this strong increase is mainly caused by the decreased electrochemical fuel oxidation rate, which manifests in an increase by 284% for the high-frequency anode activation polarization processes. Additionally, the low-frequency impedance arcs of the measured spectra show a notable change, which reveals strongly decreased gas transport of CO/CO_2 along with an increased resistance of H_2/H_2O diffusion. This observation is assigned to the poisoning of the WGS reaction, which implies a dramatically lowered utilization of the CO within the fuel.

For the first time, a separate analysis of the impact of sulfur on the single polarization processes of a SOFC is presented. Based on the findings of this work (cf. Chapters 5 and 6), the applied method facilitates a separate monitoring of the impact of sulfur poisoning on (i) the electrochemical fuel oxidation and (ii) the reforming reactions within anode supported SOFCs.

8. Further Works

8.1. EIS Modeling of Detailed Chemistry in SOFC Anodes

In a recent publication, Zhu et al. [160] presented an electrochemical cell model applied to assist understanding the effects of detailed surface chemistry on solid oxide fuel cell (SOFC) electrochemical impedance spectra. This contribution provides significant new understanding about the role of detailed heterogeneous chemistry on electrochemical impedance spectra. It is the result of a collaboration between the Colorado School of Mines (CSM), the Institute of Chemical Technology and Polymer Chemistry (ITCP) and the Institut für Werkstoffe der Elektrotechnik (IWE) (both KIT). Basis for the discussion of this model are the impedance data generated on reformate fueled SOFCs presented in this thesis. With the measurements taken under equilibrium reformate compositions (cf. Chapter 3) and the numerical accuracy of the physically motivated equivalent circuit model (ECM) fit (cf. Chapter 4), a quantitative analysis of the single-loss contributions of the measured impedance spectra for strictly defined operating conditions is possible.

The physical model, which is described in detail in Ref. [128] represents one-dimensional transport and chemistry through the thickness of a membrane-electrode structure. Porous-media transport within the electrode structure is represented by the Dusty-Gas model [170]. Charge-transfer chemistry at the electrode-electrolyte interface is modeled in a modified Butler-Volmer form (cf. Section 2.5.2). Heterogeneous (catalytic reforming) chemistry on Ni is represented by an elementary reaction mechanism [122, 123]. Detailed chemical kinetics is widely used in general catalysis research [171], a detailed introduction to its yet relatively infrequent application in SOFC modeling can be found in Ref. [172]. Table 8.1 lists the applied 21-step mechanism that represents catalytic chemistry of CH_4, H_2, and CO oxidation and reforming on the Ni surfaces.

Unlike global reactions that consider only gas-phase species, the detailed mechanisms consider both gas-phase and surface-adsorbed species. The phase is noted in parentheses, with (g) representing the gas and (Ni) representing the nickel surface. An open surface site on the Ni surface is represented as (Ni). Because detailed heterogeneous reaction mechanisms include surface-adsorbed species, the surface species introduce a capacity effect that affects the complex impedance. Such effects cannot be predicted by models that neglect heterogeneous surface chemistry entirely or use global reactions that involve

Table 8.1.: Heterogeneous reaction mechanism for CH_4 reforming on a Ni-based catalyst applied in Ref. [160].

1.	$H_2(g) + 2\,(Ni) \rightleftharpoons H(Ni) + H(Ni)$
2.	$O_2(g) + (Ni) + (Ni) \rightleftharpoons O(Ni) + O(Ni)$
3.	$CH_4(g) + (Ni) \rightleftharpoons CH_4(Ni)$
4.	$H_2O(g) + (Ni) \rightleftharpoons H_2O(Ni)$
5.	$CO_2(g) + (Ni) \rightleftharpoons CO_2(Ni)$
6.	$CO(g) + (Ni) \rightleftharpoons CO(Ni)$
7.	$O(Ni) + H(Ni) \rightleftharpoons OH(Ni) + (Ni)$
8.	$OH(Ni) + H(Ni) \rightleftharpoons H_2O(Ni) + (Ni)$
9.	$OH(Ni) + OH(Ni) \rightleftharpoons O(Ni) + H_2O(Ni)$
10.	$O(Ni) + C(Ni) \rightleftharpoons CO(Ni) + (Ni)$
11.	$O(Ni) + CO(Ni) \rightleftharpoons CO_2(Ni) + (Ni)$
12.	$HCO(Ni) + (Ni) \rightleftharpoons CO(Ni) + H(Ni)$
13.	$HCO(Ni) + (Ni) \rightleftharpoons O(Ni) + CH(Ni)$
14.	$CH_4(Ni) + (Ni) \rightleftharpoons CH_3(Ni) + H(Ni)$
15.	$CH_3(Ni) + (Ni) \rightleftharpoons CH_2(Ni) + H(Ni)$
16.	$CH_2(Ni) + (Ni) \rightleftharpoons CH(Ni) + H(Ni)$
17.	$CH(Ni) + (Ni) \rightleftharpoons C(Ni) + H(Ni)$
18.	$O(Ni) + CH_4(Ni) \rightleftharpoons CH_3(Ni) + OH(Ni)$
19.	$O(Ni) + CH_3(Ni) \rightleftharpoons CH_2(Ni) + OH(Ni)$
20.	$O(Ni) + CH_2(Ni) \rightleftharpoons CH(Ni) + OH(Ni)$
21.	$O(Ni) + CH(Ni) \rightleftharpoons C(Ni) + OH(Ni)$

only gas-phase species. Figure 8.1 illustrates the capacitive effect of surface chemistry on the predicted impedance spectra for a variation of the specific surface area within the electrode structure.

By increasing the specific surface area, the peak frequency can be significantly reduced. It has to be noted that increasing the surface area has negligible influence on the gas diffusion resistance. For H_2/H_2O mixtures, the observed predictions can be interpreted as storage of fuel gas species on the Ni surface, thus introducing a mainly capacitive effect to the transient concentration polarization.

Comparison of the model-predicted Electrochemical Impedance Spectroscopy (EIS) responses with an extensive set of experimental measurements shows that the model predicts all the qualitative trends that are observed. Figure 8.2 depicts the comparison of predicted and measured EIS spectra for H_2/H_2O mixtures.

The experimental spectra were obtained by extracting the reaction-diffusion behavior (i.e., the general finite-length Warburg (GFLW) process P_{1A}) from the complex nonlinear least squares (CNLS)-fit of the ECM as presented in Section 4.5. The excellent agreement between modeled and measured area specific resistance (ASR) shows that the model accurately represents the polarization process P_{1A}. Only for very low H_2O levels, the simulation shows slightly higher resistances than the measurement. Also the predicted peak frequencies are close to the experimental measurements. An interesting fact however,

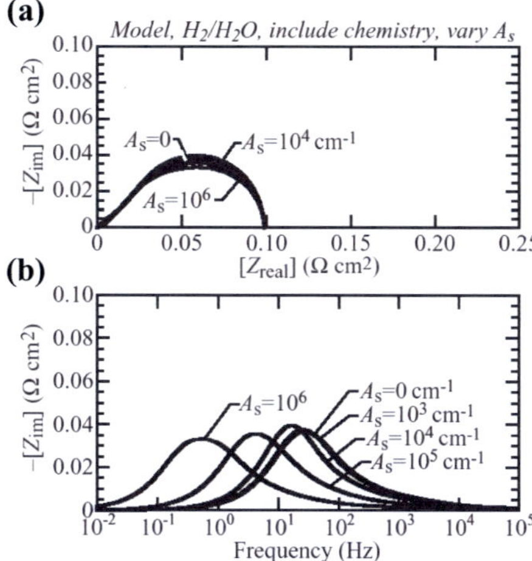

Figure 8.1.: Predicted reaction-diffusion impedance for a fuel mixture of $15\,\%$ H_2, $12.5\,\%$ H_2O and $72.5\,\%$ N_2 by varying the specific surface area. (a) Nyquist representation and (b) imaginary components of the complex impedances as functions of frequency. [160]

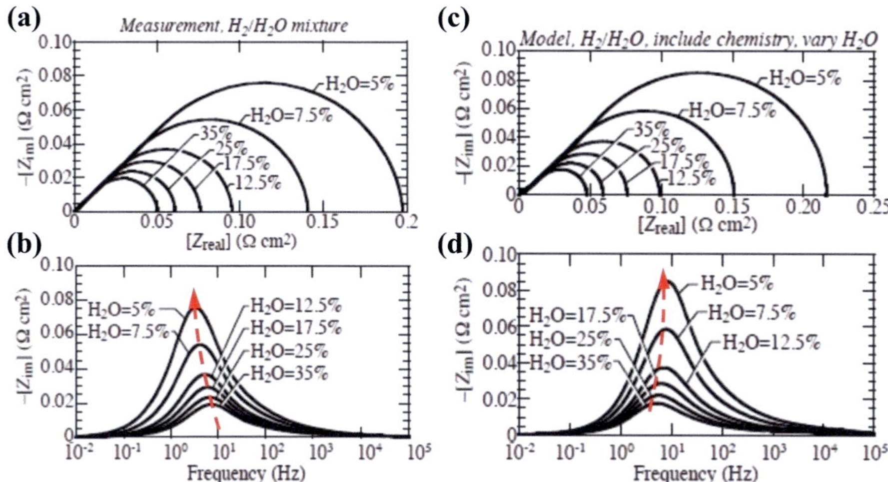

Figure 8.2.: Nyquist plots and frequency-dependent imaginary parts of the complex anode transport impedance for varied $p\mathrm{H_2O_{an}}$ in the H_2/H_2O anode fuel mixture. (a,b) Measured impedance data and (c,d) model predictions (fuel compositions see Table 5.1; cathode: air; $T = 800\,^\circ\mathrm{C}$). [cell# Z2_190], [160]

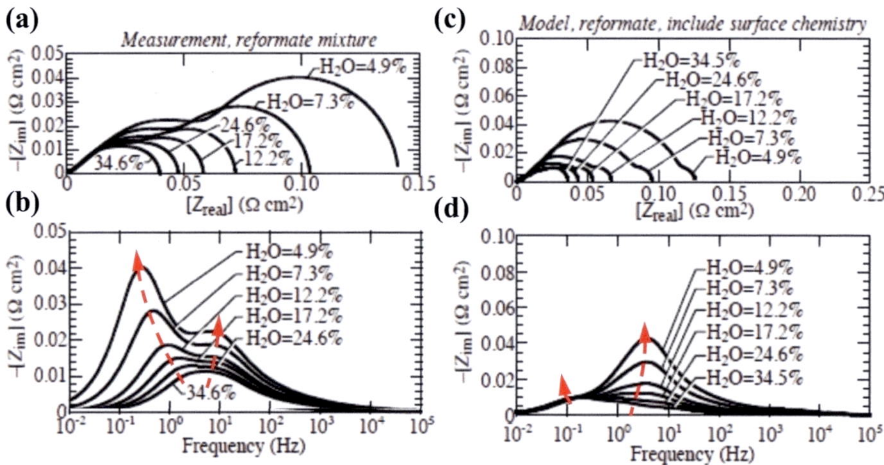

Figure 8.3.: Nyquist plots and frequency-dependent imaginary parts of the complex anode transport impedance for varied $p\text{H}_2\text{O}_{\text{an}}$ in the model reformate fuel mixture. (a,b) Measured impedance data and (c,d) model predictions (fuel compositions see Table 5.1; cathode: air; $T = 800\,^{\circ}\text{C}$). [cell# Z2_190], [160]

which is revealed by the comparison is that the evolution of the characteristic frequencies exhibit different trends for measurement and simulation. In contrast to the experiments (Fig. 5), it is noted that the model predicts a slight shift to higher peak frequencies as H$_2$O decreases. This shows that further investigations are necessary to fully understand the role of the detailed surface kinetics. It would be very interesting to (i) identify the determining rate expression and (ii) perform a sensitivity analysis on the relevant kinetic rate expressions according to the experimental data. With this method, it is possible to identify detailed elementary kinetic data.

Figure 8.3 depicts the comparison of predicted and measured EIS spectra for operation on a model reformate. The experimental spectra were obtained by extracting the reaction-diffusion behavior (i.e., $P_{1A} + P_{ref}$) from the CNLS-fit of the ECM as presented in Section 4.5. Also here, the ASR of the model predictions show good overall agreement with the measured EIS spectra. Furthermore, it is clearly observable that the model is capable to qualitatively reproduce the observed impedance response with two characteristic time constants. The contribution of the two polarization processes to the overall spectra as well as the evolution of the characteristic frequencies shows different behavior for measurements and model predictions. This is most probably due to inaccuracies in the underlying elementary kinetic data, which in the context of global reforming reactions like the steam reforming reaction of methane are of minor subordinate relevance, in this context however likely to produce deviations in time constants. This shows again, that it would be very interesting to employ the method of EIS measurements and simulation in order to achieve insights into detailed elementary kinetic data for heterogeneous catalytic chemistry on Ni surfaces.

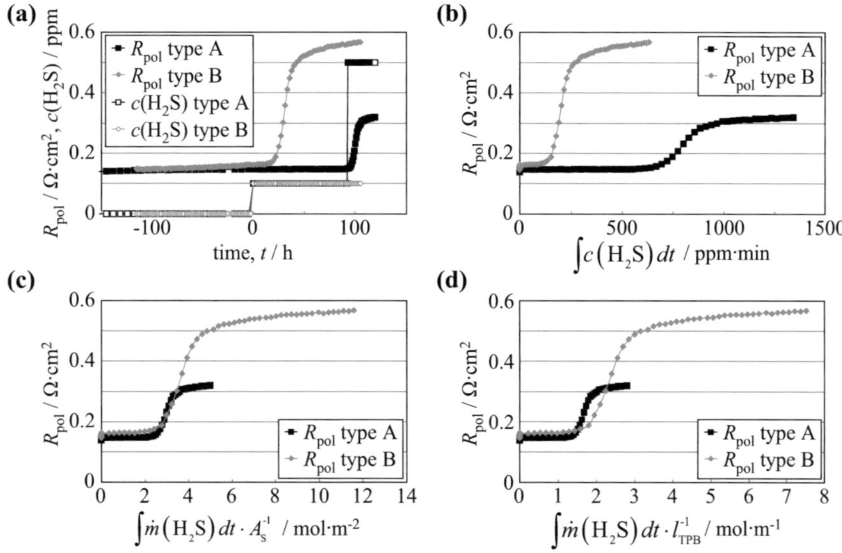

Figure 8.4.: Evolution of the polarization resistance R_{pol} plotted over (a) time, (b) H_2S-concentration times operating time, (c) total H_2S-amount divided by the Ni-surface area and (d) total H_2S-amount divided by the TPB-length. [173]

8.2. H_2S Accumulation on Ni Surfaces in SOFC Anodes

The investigations on sulfur poisoning on SOFC anodes presented in Chapter 7 of this thesis initiated a series of further studies on this topic at the IWE. The recent investigation of Weber et al. [173] on the mechanisms of sulfur poisoning in SOFC anodes is built on the results presented in Chapter 7. In this contribution, the time-resolved clear separation of the impact of sulfur poisoning on (i) the electrochemical fuel oxidation and on (ii) the reforming chemistry is used to investigate the dynamics of the sulfur coverage on Ni/YSZ.

To obtain information about the poisoning mechanism, the temporal characteristics of the ASR-increase is compared for two cells with Ni/YSZ anodes, differing in anode microstructure (type A and type B).

Figure 8.4 depicts the comparison, where the ASR-values of the polarization resistance are plotted over

a) time,

b) concentration of H_2S multiplied by operating time with H_2S-containing reformate,

c) accumulated amount of H_2S supplied to the cell divided by the Ni-surface area available in the anode substrate and the anode functional layer and

d) accumulated amount of H_2S supplied to the cell divided by the three phase boundary (TPB)-length in the anode substrate and the anode functional layer (AFL).

With this method, it could be demonstrated clearly that the accumulated H_2S-amount per Ni-surface area inside the anode substrate and AFL determines the onset of the degradation. Only from this perspective, a good agreement is achieved between the two different cells (Fig. 8.4).

9. Summary

The fuel flexibility is a very attractive feature of solid oxide fuel cells (SOFCs). Due to the high operating temperatures and the catalytic activity of the anode materials, SOFCs can be operated with hydrocarbon fuels. Theoretically, the SOFC is capable to directly convert the chemical energy of the fuels into electric energy. Apart from achieving long-term stability, it is one of the most important development goals for the SOFC to understand the electrochemical loss mechanisms that determine the performance of the single cells. For the operation with hydrocarbon fuels, there is a great deal of knowledge about the conversion of the fuels at SOFC anodes in terms of catalytic reforming chemistry, whereas little is known about the electrochemistry of the cells. No overall understanding of the actual electrochemical processes that determine the electrical performance of the cells is available in literature.

It is aim of this work to capture the electrochemical loss processes of the SOFC for operation with hydrocarbon fuels. In particular, this comprises the identification of (i) the electrochemical fuel oxidation mechanism, (ii) the fuel gas transport properties in porous electrode structures and (iii) the interaction of heterogeneous reforming chemistry with the above mentioned processes.

Present electrochemical analyses are complicated by uncertainties in temperature and species concentration at the anode. This is due to the highly endothermic reforming reactions, which apart from converting the fuel gas species, induce strong temperature gradients at the cell. In this thesis, electrochemical analyses were performed under operation on model reformates. The applied model fuels represent fuel gas mixtures of H_2, H_2O, CO, CO_2 and N_2 at chemical equilibrium. With this method, species conversion via the reforming reactions is inhibited and analyses under strictly defined operating parameters are guaranteed. The analyses were performed via Electrochemical Impedance Spectroscopy (EIS), a small-signal measuring technique, under which the processes at the anode are minimally deviated from their chemical equilibria. Hence, no uncertainties in operating parameters are induced by the chosen measuring technique.

In theory, EIS offers the advantage of separating the electrochemical processes that form the inner resistance of SOFC. This measuring technique makes use of the fact that the individual electrochemical processes of the SOFC exhibit different characteristic time constants. By periodically stimulating the cell at different relaxation frequencies, the individual electrochemical processes give differing contributions to the overall response of the system. In the measured electrochemical impedance spectra, information about the single contributions of the individual electrochemical loss processes to the overall cell resistance is provided. In practice, the overlapping time scales of the individual loss processes of the SOFC inhibit a separated analysis directly from the measured spectra. In this work, the application of the distribution of relaxation times (DRT), an impedance analysis method developed at Institut für Werkstoffe der Elektrotechnik (IWE)[1] enables the separation of the single loss processes from the EIS spectra measured on the real cells.

The investigations were performed on state of the art anode supported cells (ASCs) manufactured by Forschungszentrum Jülich (FZJ), one of the best-performing, most stable and hence most investigated SOFC systems available. The cells were built on cermet anode structures with a $1000\ \mu m$ anode substrate and an anode functional layer (AFL) with a thickness of several μm. The cermet anodes were made of nickel (Ni) and yttria stabilized zirconia (YSZ), which is the most common SOFC anode material system.

In order to identify the electrochemical loss processes for the reformate operation of these cells, EIS analyses were performed under systematic variation of the operating parameters. The analysis over a broad range of operating temperatures and H/C/O-ratios of the fuel gas mixture lead to the identification and physical assignment of four different anode-side loss processes. Next to the processes assigned to the gas diffusion in the anode substrate (P_{1A}) and the coupling of anodic charge transfer reaction and the ionic conduction in the AFL (P_{2A}, P_{3A}) an additional electrochemical process (P_{ref}), never reported before, with a relaxation frequency below 1 Hz has been identified. The newly identified low-frequency process P_{ref} shows the characteristics of gas transport processes, which are probably linked to reforming chemistry.

From this knowledge, a physically meaningful equivalent circuit model (ECM) was set up, representing the impedance response of the investigated cells under reformate operation. In the ECM, each electrochemical process is represented by a single equivalent circuit element. complex nonlinear least squares (CNLS)-fit of the ECM to measured impedance spectra facilitate a separated, quantitative analysis of the individual electrochemical loss processes at the anode.

This method has been applied in order to understand the electrochemical fuel oxidation mechanism for the reformate fueled anode in a kinetic study. To this end, the anode activation polarization processes were compared for reformate operation and operation on hydrogen (i.e., fuel gas mixtures of H_2 and H_2O). For a variation of H_2 and H_2O concentrations at the anode as well as the operating temperature of the cell, the processes P_{2A} and P_{3A} showed the same qualitative behavior for both fuel types. From the anode activation polarization resistance ($R_{act,an} = R_{2A} + R_{3A}$) obtained with the help of the CNLS-fit, activation energy of the electrochemical fuel oxidation and reaction orders with respect to the involved species were determined. The kinetic parameters are identical for operation on reformate and hydrogen, but considerably different for the electrochemical oxidation of CO. It is hence unambiguously demonstrated that for the reformate fueled Ni/YSZ anodes, exclusively H_2 is electrochemically oxidized. This finding further implicates that other electrochemically active species in the reformate fuel are subsequently thermally oxidized via reforming chemistry. For CO, thermal oxidation proceeds namely via the water-gas shift (WGS) reaction.

---

[1]PhD Dissertation Helge Schichlein, Ref. [65].

The gas transport pathways for the reactants of the electrochemical fuel oxidation were investigated by means of a combined EIS and finite element method (FEM) modeling analysis. Based on the identification of the electrochemical fuel oxidation mechanism, a schematic model understanding of the gas transport properties in reformate fueled Ni/YSZ anodes was developed. Porous-media gas transport in the electrode structure must accommodate two gas transport pathways (H_2/H_2O and CO/CO_2), which are coupled via the WGS reaction. In this study, it was demonstrated that the impedance response of this coupled diffusion-reaction scheme represents the two low-frequency anode polarization processes P_{1A} and P_{ref}. A further analysis of the underlying transport processes revealed that the faster process P_{1A} can be assigned to the diffusion of H_2/H_2O, whereas the low-frequency process P_{ref} is dominated by CO/CO_2 diffusion.

For the Ni/YSZ anodes applied herein, (i) the electrochemical fuel oxidation mechanism and (ii) the gas transport properties in the porous electrode structure were identified by means of EIS. The results comprise an overall understanding of the electrochemical loss processes of reformate fueled SOFC anodes, which has not been presented in literature before.

As a last point, the developed methods were applied in a poisoning study investigating the impact of sulfur contaminants in reformate fuels on the performance of the SOFC. For the first time in literature, the impact of H_2S on the individual electrochemical loss processes was analyzed. Under operation with H_2S containing fuels, the anode activation polarization processes P_{2A} and P_{3A} exhibit a strong degradation. At the low-frequency processes, a strong increase of P_{1A} along with a significant decrease of P_{ref} was observed. With the findings discussed above, the observed degradation of the individual loss processes could be assigned to two different poisoning mechanisms. The degradation of the anode activation polarization processes P_{2A} and P_{3A} is ascribed to the poisoning of the electrochemical fuel oxidation. The contrary development of the gas transport processes P_{1A} and P_{ref} can be traced back to the poisoning of the catalytic reforming chemistry inside the Ni/YSZ electrode structure.

Sulfur-poisoning of the electrochemistry implicates an increase in the inner resistance, which is a decrease of the cell performance. The observed sulfur-poisoning of the reforming chemistry has a minor impact on the decrease in cell resistance, it implicates however a decreased ability of the SOFC to utilize parts of the fuel provided. The former issue demands development of sulfur tolerant anode materials, while the latter can be solved by application of a sulfur tolerant shift catalyst in the fuel gas channel. It is demonstrated that the developed measuring technique along with the findings of this thesis can assist in analyzing the success of further development on both aspects.

Appendix

A. Applicability to Further Systems

As an outlook, results on metal supported SOFCs are presented. The single cells were developed in the joint project METSOFC,[2] funded by the European Union (EU) 7th framework programme and electrochemically characterized at IWE. The anode of the cells consist of a metal backbone, a dip-coated gadolinium doped ceria (GDC) layer and infiltrated electrocatalysts. Details regarding the cell manufacturing, performance and electrochemical loss processes can be found in Refs. [45, 46].

Table A.1.: Anodic partial pressures of H_2, H_2O, CO, CO_2 and N_2 selected for the EIS measurements for varying shares of H_2/H_2O and CO/CO_2 in the anode fuel gas of the metal supported cell.

CO/CO_2-share (%)	$pH_{2,\text{an}}$ (atm)	pH_2O_{an} (atm)	pCO_{an} (atm)	$pCO_{2,\text{an}}$ (atm)
0	0.40	0.60	0.00	0.00
20	0.32	0.48	0.08	0.12
40	0.24	0.36	0.16	0.24
60	0.16	0.24	0.24	0.36
80	0.08	0.12	0.32	0.48
90	0.04	0.06	0.36	0.54
95	0.02	0.03	0.38	0.57
100	0.00	0.00	0.40	0.60

The cell has been tested for various reformate fuels. Starting from pure H_2/H_2O, parts of the fuel were replaced by CO/CO_2 in various steps until the share of CO/CO_2 reached 100%. The total anodic gas flow was kept constant at a value of $250\,\text{sccm}$ and the cell was operated at a simulated fuel utilization of 60%. The detailed gas compositions of the applied fuel gas flow are listed in Table A. It has to be noted that the applied gas mixtures were not in chemical equilibrium. At the operating temperature of $650\,^{\circ}\text{C}$, the actual gas compositions at the cell should be swifted somewhat more to H_2/CO_2 (cf. Table 4.4 for comparison). At each step, EIS spectra were recorded.

The DRT for the impedance spectra recorded at each fuel composition are given in Fig. A.1. With increasing content of CO/CO_2 within the fuel, (i) the peaks attributed

---

[2]http://www.metsofc.eu

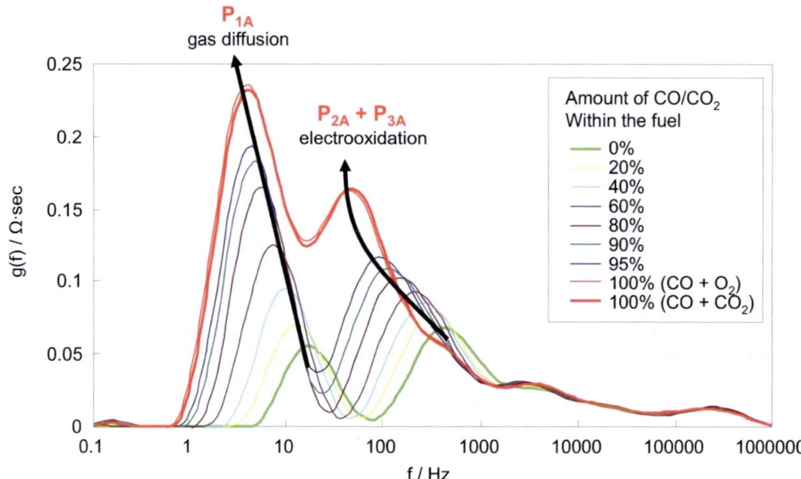

Figure A.1.: DRT calculated from EIS spectra recorded for varying shares of H_2/H_2O and CO/CO_2 in the anode fuel gas (fuel composition see Table A; cathode: air; $T = 650\,°C$). [cell# Z7_127]

to the gas diffusion through the metal support and (ii) the peaks attributed to the electrochemical oxidation of the fuel within the anode are rising. This refers to an increase of the corresponding polarization resistances. The measurements for $100\,\%$ CO/CO_2 demonstrate that the anode is able to electrochemically oxidize CO.

The pronounced gap between the curves for $95\,\%$ and $100\,\%$ CO/CO_2 is similar to the one observed for Ni/YSZ anodes (cf. Fig. 4.5). It indicates that also for reformate fueled METSOFC anodes, H_2 is preferentially electrochemically oxidized. However has to be noted that in contrast to the results on Ni/YSZ anodes, a stronger increase of the activation polarization processes with rising CO/CO_2 is visible here. An exclusive electrochemical oxidation of H_2 can thus not be determined with these preliminary results. A kinetic analysis, as demonstrated in Chapter 5, should lead to the clear identification of the fuel oxidation mechanism.

The multi-arc impedance response of the diffusion-reaction processes, as reported for Ni/YSZ anodes in this thesis (cf. Chapters 4 and 6), is not observed for the present METSOFC anode. The preferred electrochemical oxidation of H_2, should yet lead to a similar gas transport scheme (H_2/H_2O and CO/CO_2 diffusion pathways coupled by the WGS reaction), as reported for Ni/YSZ anodes in Chapter 6. The appearance of only one peak in the DRT can be explained by (i) the strong overlap of the relaxation frequencies of H_2/H_2O and CO/CO_2 diffusion in the highly porous metal substrates and (ii) further uncertainties caused by the fact that the applied fuel gas composition was not in chemical equilibrium (cf. Section 2.6.2). The continuous increase of this peak from operation on pure H_2/H_2O to pure CO/CO_2 however demonstrates, as in the case of Ni/YSZ (cf. Fig. 4.5), that the reforming chemistry does not appear as a single polarization process in the measured spectra. A poisoning analysis as performed in Chapter 7 should deliver more insights on the detailed gas transport properties of the reformate fueled metal-supported anode.

B. Gibbs Diagrams

For operation with reformate fuels (mixtures of H_2, H_2O, CO, CO_2 and N_2), it is safe to assign the likelihood of carbon deposition by determining the formation of carbon in the equilibrium gas composition of the applied fuel. In this section, the ternary C-H-O phase diagrams (Gibbs triangles) calculated with MALT [152] are shown for the reformate fuel gas compositions applied in this work.

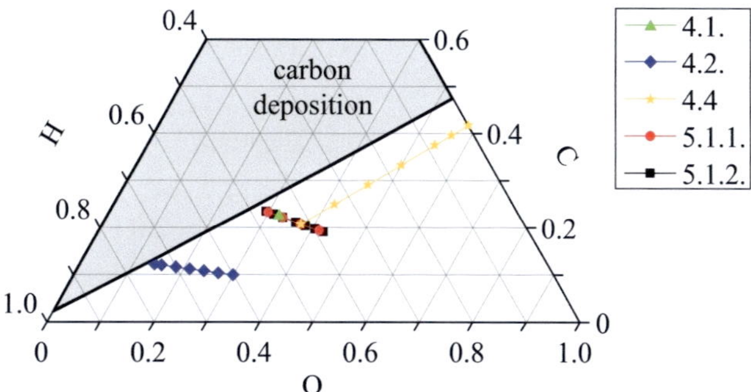

Figure B.2.: Equilibrium C-H-O diagram showing the carbon isotherm (solid line) at 800 °C and 1.0 atm for graphite and amorphous carbons. The data points correspond to the compositions of the performed fuel gas variations in Sections 4.1., 4.2., 4.4., 5.1.1. and 5.1.2., respectively.

Figure B.2 depicts the equilibrium C-H-O diagram for 800 °C and 1.0 atm. The black line indicates the carbon isotherm for graphite and amorphous carbons. The data points give the C-H-O ratio of the equilibrium compositions of the performed fuel gas variations at 800 °C (cf. Sections 4.1, 4.2, 4.4, 5.1.1 and 5.1.2). All measuring points lie below the isotherms for carbon formation, which means that carbon formation is not expected from thermodynamics at these points.

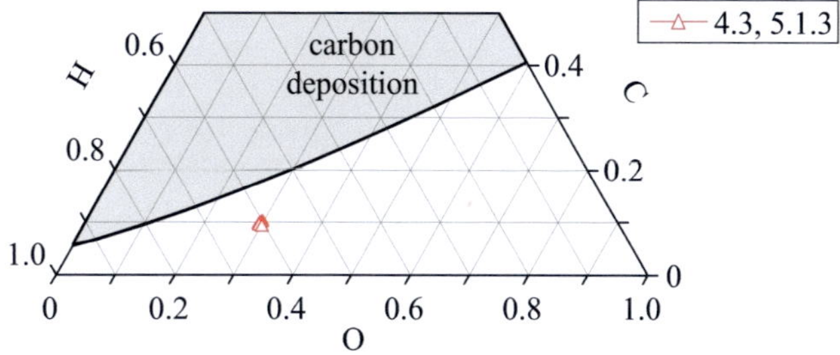

Figure B.3.: Equilibrium C-H-O diagram showing the carbon isotherm (solid line) at 680 °C and 1.0 atm for graphite and amorphous carbons. The data points correspond to the fuel gas compositions for the performed temperature variation discussed in Sections 4.3. and 5.1.3.

For the measurements performed under variation of operating temperature (cf. Sections 4.3 and 5.1.3), it is sufficient to analyze the carbon isotherm for the lowest operating temperature. Accordingly, the equilibrium C-H-O diagram for 680 °C and 1.0 atm is depicted in Fig. B.3. Also here, all measuring points lie below the isotherms for carbon formation, indicating that carbon formation is not expected at these points.

C. Measured Samples

Table C.2.: List of the tested single cells presented in this thesis.

IWE-ID	supplier ID	supplier	performed measurements
Z2_188	10905	FZJ	variation of fuel humidification (cf. Chapter 4)
Z2_190	10536	FZJ	variation of T, $p\mathrm{H}_{2,\mathrm{an}}$ and $p\mathrm{H}_2\mathrm{O}_\mathrm{an}$ (cf. Chapters 4 and 5)
Z2_211	15358	FZJ	variation of reformate composition (cf. Chapter 4)
Z6_127	10540	FZJ	H_2S poisoning (cf. Chapter 7)
Z7_127	SM2922#2	Risø-DTU	metal supported cell (cf. App. A)

D. Supervised Diploma Theses and Study Projects

- Sebastian Hirn: "3D-Modellierung und Simulation der Stofftransportvorgänge im Brenngas (H_2:H_2O) der Hochtemperatur-Brennstoffzelle". Study project, 2011.

- Sebastian Dierickx: "Schwefelvergiftung im Reformatbetrieb der Hochtemperatur-brennstoffzelle SOFC". Bachelor thesis (B.Sc.), 2011.

- Helge Geisler: "Transiente Simulation der Stofftransportvorgänge im Anodensubstrat der SOFC". Diploma thesis (Dipl.-Ing.), 2011.

- Christoph Müller and Michael Rößler: "Schwefelvergiftung an SOFC-Anoden". Bachelor thesis (B.Sc.), 2012.

- Blasius Forreiter: "Einfluss von neuartigen Herstellungsverfahren auf die elektrochemischen Eigenschaften der SOFC". Bachelor thesis (B.Sc.), 2012.

- Sebastian Hirn: "Transiente Simulation der Stofftransportvorgänge im Brenngas der Hochtemperatur-Brennstoffzelle SOFC" (Co-supervised with H. Geisler). Diploma thesis (Dipl.-Ing.), 2012.

E. Publications

- A. Kromp, J. Nielsen, P. Blennow, T. Klemensø and A. Weber, "Break-down of Losses in High Performing Metal-Supported Solid Oxide Fuel Cells", *Fuel Cells*, published online, DOI: 10.1002/fuce.201200165 (2013).

- A. Weber, S. Dierickx, A. Kromp and E. Ivers-Tiffée, "Sulphur Poisoning of Anode-Supported SOFCs under Reformate Operation", *Fuel Cells*, published online, DOI: 10.1002/fuce.201200180 (2013).

- J.-C. Njodzefon, D. Klotz, A. Kromp, A. Weber and E. Ivers-Tiffée, "Electrochemical Modeling of the Current-Voltage Characteristics of an SOFC in Fuel Cell and Electrolyzer Operation Modes", *Journal of the Electrochemical Society* 160 (4), pp. F313-F323 (2013).

- H. Zhu, A. Kromp, A. Leonide, E. Ivers-Tiffée, O. Deutschmann and R. J. Kee, "A Model-Based Interpretation of the Influence of Anode Surface Chemistry on Solid Oxide Fuel Cell Electrochemical Impedance Spectra", *Journal of the Electrochemical Society* 159 (7), pp. F255-F266 (2012).

- B. J. McKenna, N. Christiansen, R. Schauperl, P. Prenninger, J. Nielsen, P. Blennow, T. Klemensø, S. Ramousse, A. Kromp and A. Weber, "Advances in Metal Supported Cells in the METSOFC EU Consortium", in F. Lefebvre-Joud (Ed.), *Proceedings of the 10th European Solid Oxide Fuel Cell Forum*, Chapter 07, pp. 20-29 (2012).

- A. Kromp, J. Nielsen, P. Blennow, T. Klemensø and A. Weber, "Break-down of Losses in High Performing Metal-Supported Solid Oxide Fuel Cells", in F. Lefebvre-Joud (Ed.), *Proceedings of the 10th European Solid Oxide Fuel Cell Forum*, Chapter 07, pp. 84-94 (2012).

- A. Weber, S. Dierickx, A. Kromp and E. Ivers-Tiffée, "Sulphur Poisoning of Anode-Supported SOFCs under Reformate Operation", in F. Lefebvre-Joud (Ed.), *Proceedings of the 10th European Solid Oxide Fuel Cell Forum*, Chapter 08, pp. 72-82 (2012).

- A. Kromp, H. Geisler, A. Weber and E. Ivers-Tiffée, "Electrochemical Impedance Modeling of Reformate-Fuelled Anode-Supported SOFC", in F. Lefebvre-Joud (Ed.), *Proceedings of the 10th European Solid Oxide Fuel Cell Forum*, Chapter 17, pp. 92-101 (2012).

- A. Kromp, A. Leonide, A. Weber and E. Ivers-Tiffée, "Electrochemistry of Reformate-Fuelled Anode-Supported SOFC", in F. Lefebvre-Joud (Ed.), *Proceedings of the 10th European Solid Oxide Fuel Cell Forum*, Chapter 18, pp. 6-13 (2012).

- A. Kromp, S. Dierickx, A. Leonide, A. Weber and E. Ivers-Tiffée, "Electrochemical Analysis of Sulfur-Poisoning in Anode Supported SOFCs Fuelled with a Model Reformate", *Journal of the Electrochemical Society* 159 (5), pp. B597-B601 (2012).

- D. Klotz, J. P. Schmidt, A. Kromp, A. Weber and E. Ivers-Tiffée, "The Distribution of Relaxation Times as Beneficial Tool for Equivalent Circuit Modeling of Fuel Cells and Batteries", *ECS Transactions* 41 (28), pp. 25-33 (2012).

- J.-C. Njodzefon, D. Klotz, A. Leonide, A. Kromp, A. Weber and E. Ivers-Tiffée, "Electrochemical Studies on Anode Supported Solid Oxide Electrolyzer Cells", *ECS Transactions* 41 (33), pp. 113-122 (2012).

- A. Kromp, S. Dierickx, A. Leonide, A. Weber and E. Ivers-Tiffée, "Electrochemical Analysis of Sulphur-Poisoning in Anode-Supported SOFCs under Reformate Operation", *ECS Transactions* 41 (33), pp. 161-169 (2012).

- A. Kromp, A. Leonide, A. Weber and E. Ivers-Tiffée, "Electrochemical Analysis of Reformate-Fuelled Anode Supported SOFC", *Journal of the Electrochemical Society* 158 (8), pp. B980-B986 (2011).

- A. Kromp, A. Leonide, A. Weber and E. Ivers-Tiffée, "Hydrogen Oxidation Kinetics in Reformate-Fuelled Anode Supported SOFC", *ECS Transactions* 35, pp. 665-678 (2011).

- P. Blennow, J. Hjelm, T. Klemensø, S. Ramousse, A. Kromp, A. Leonide and A. Weber, "Manufacturing and characterization of metal-supported solid oxide fuel cells", *Journal of Power Sources* 196, pp. 7117-7125 (2011).

- A. Kromp, A. Leonide, H. Timmermann, A. Weber and E. Ivers-Tiffée, "FEM-Modeling of Internal Reforming in SOFC-Anodes", in DECHEMA (Ed.), *First International Conference on Materials for Energy, Extended Abstracts - Book A*, pp. 114-116 (2010).

- A. Kromp, A. Leonide, H. Timmermann, A. Weber and E. Ivers-Tiffée, "Internal Reforming Kinetics in SOFC-Anodes", *ECS Transactions* 28, pp. 205-215 (2010).

- P. Blennow, J. Hjelm, T. Klemensø, S. Ramousse, A. Kromp, A. Leonide and A. Weber, "Manufacturing and Characterization of Metal Supported SOFCs", in J. T. S. Irvine and U. Bossel (Eds.), *Proceedings of the 9th European Solid Oxide Fuel Cell Forum*, pp. 16-1-16-19 (2010).

- A. Kromp, A. Leonide, H. Timmermann, A. Weber and E. Ivers-Tiffée, "Internal Reforming Kinetics in SOFC-Anodes", in J. T. S. Irvine and U. Bossel (Eds.), *Proceedings of the 9th European Solid Oxide Fuel Cell Forum*, pp. 6-103-6-110 (2010).

F. Conference Contributions

- A. Kromp, S. Dierickx, A. Leonide, A. Weber and E. Ivers-Tiffée, Oral presentation by A. Kromp: "Sulfur-Poisoning in Reformate Fuelled Anode Supported Solid Oxide Fuel Cells", 222th Meeting of The Electrochemical Society (Honolulu, USA), 07.10. - 12.10.2012.

- B. J. McKenna, N. Christiansen, R. Schauperl, P. Prenninger, J. Nielsen, P. Blennow, T. Klemensø, S. Ramousse, A. Kromp and A. Weber, Oral presentation by B. J. McKenna: "Advances in Metal Supported Cells in the METSOFC EU Consortium", 10th European SOFC Forum (Lucerne, Switzerland), 26.06. - 29.06.2012.

- A. Kromp, J. Nielsen, P. Blennow, T. Klemensø and A. Weber, Poster presentation by A. Kromp: "Break-down of Losses in High Performing Metal-Supported Solid Oxide Fuel Cells", 10th European SOFC Forum (Lucerne, Switzerland), 26.06. - 29.06.2012.

- A. Weber, S. Dierickx, A. Kromp and E. Ivers-Tiffée, Poster presentation by A. Weber: "Sulphur Poisoning of Anode-Supported SOFCs under Reformate Operation", 10th European SOFC Forum (Lucerne, Switzerland), 26.06. - 29.06.2012.

- A. Kromp, H. Geisler, A. Weber and E. Ivers-Tiffée, Poster presentation by A. Kromp: "Electrochemical Impedance Modeling of Reformate-Fuelled Anode-Supported SOFC", 10th European SOFC Forum (Lucerne, Switzerland), 26.06. - 29.06.2012.

- A. Kromp, A. Leonide, A. Weber and E. Ivers-Tiffée, Oral presentation by A. Kromp: "Electrochemistry of Reformate-Fuelled Anode-Supported SOFC", 10th European SOFC Forum (Lucerne, Switzerland), 26.06. - 29.06.2012.

- A. Kromp, A. Leonide, A. Weber and E. Ivers-Tiffée, Oral presentation by A. Kromp: "Electrochemical Analysis of Reformate-Fuelled Anode-Supported SOFC", 220th Meeting of The Electrochemical Society (Boston, USA), 09.10. - 14.10.2011.

- A. Kromp, A. Leonide, A. Weber and E. Ivers-Tiffée, Oral presentation by A. Kromp: "Performance Analysis of Anode-Supported SOFC under Reformate-Operation", 62nd Annual ISE Meeting (Niigata, Japan), 11.09. - 16.09.2011.

- A. Kromp, A. Leonide, A. Weber and E. Ivers-Tiffée, Poster presentation by A. Kromp: "Hydrogen-Oxidation Kinetics in Reformate Fuelled Anode Supported SOFC", SOFC XII (Montréal, Canada), 01.05. - 06.05.2011.

- A. Kromp, A. Leonide, A. Weber and E. Ivers-Tiffée, Oral presentation by A. Kromp: "Performance of Solid Oxide Fuel Cells Operated with a Model Reformate Gas", 218th Meeting of The Electrochemical Society (Las Vegas, USA), 10.10. - 15.10.2010.

- A. Kromp, A. Leonide, H. Timmermann, A. Weber and E. Ivers-Tiffée, Poster presentation by A. Kromp: "Modeling Internal Reforming Kinetics in SOFC-Anodes", 61st Annual ISE Meeting (Nice, France), 26.09. - 01.10.2010.

- A. Kromp, A. Leonide, H. Timmermann, A. Weber and E. Ivers-Tiffée, Poster presentation by A. Kromp: "FEM-Modeling of Internal Reforming Kinetics in SOFC-Anodes", First International Conference on Materials for Energy (Karlsruhe, Germany), 04.07. - 08.07.2010.

- P. Blennow, T. Klemensø, S. Ramousse, A. Kromp, A. Leonide and A. Weber, Oral presentation by P. Blennow: "Manufacturing and Characterization of Metal Supported SOFCs", 9th European SOFC Forum (Lucerne, Switzerland), 29.06. - 02.07.2010.

- A. Kromp, A. Leonide, H. Timmermann, A. Weber and E. Ivers-Tiffée, Poster presentation by A. Kromp: "Internal Reforming Kinetics in SOFC-Anodes", 9th European SOFC Forum (Lucerne, Switzerland), 29.06. - 02.07.2010

- A. Kromp, A. Leonide, H. Timmermann, A. Webera and E. Ivers-Tiffée, Poster presentation by A. Kromp: "Internal Reforming Kinetics in SOFC-Anodes", 217th Meeting of The Electrochemical Society (Vancouver, Canada), 25.04. - 30.04.2010

G. Symbols

Symbol	Description	Unit/Value
A_s	catalyst-specific surface area	1/m
a	reaction order of H_2	-
$\mathbf{a}$	fit parameter vector	-
b	reaction order of H_2O	-
c	reaction order of O_2	-
c_k	molar concentration of species k	mol/m^3
D_k	diffusion coefficient	m^2/s
D_k^eff	effective diffusion coefficient	m^2/s
$D_{\mathrm{kn},k}$	Knudsen diffusion coefficient	m^2/s

Symbol	Description	Unit/Value
$D_{\mathrm{mol},k}$	molar diffusion coefficient	m^2/s
D^δ	chemical diffusion coefficient	m/s
E_{act}	activation energy	eV or J/mol
E_{rel}	residual	%
F	Faraday constant	96500 C/mol
f	frequency	Hz
f_{c}	characteristic frequency	Hz
$g(f)$	distribution function (numerical)	Ωs
h_{g}	half-width relative to peak height	-
I	static current	A
$i(t)$	transient current density	A/cm^2
$i_{0,el}$	electrode exchange current density	A/m^2
$J_{\mathrm{d},k}$	diffusion flux	mol/m^{2}s
j	imaginary unit	$\sqrt{-1}$
K_r	equilibrium constant of reaction r	-
k_r	velocity constant	mol/m^{3}Pa^{2}s
k^δ	chemical surface exchange coefficient	m/s
L	geometric length	m
l_{el}	electrode thickness	m
M_k	molar mass	kg/mol
P	conversion factor	101330 Pa/atm
P_0	absolute pressure	1 atm
P_i	polarization process i	-
p_k	partial pressure	atm
R	ideal gas constant	8.314 J/molK
$R_{\mathrm{pol},i}$	polarization resistance	Ω
R_r	reaction rate	kg/m^{3}s
T	temperature	K
t	time	s
U	voltage	V
$u(t)$	transient voltage	V
w	weighting factor	-
x	direction of the electrode thickness	m
x_k	mole fraction	-
x_k^{GC}	mole fraction in the gas channel	-
x_k^{TPB}	mole fraction at the TPB	-
$\underline{Z}$	complex impedance	Ω or Ωcm^2
Z'	real part of the impedance	Ω or Ωcm^2
Z''	imaginary part of the impedance	Ω or Ωcm^2
z	number of transferred electrons	z
ΔG	Gibbs free energy	J/mol
ΔH°	standard enthalpy of reaction	J/mol
Δp_k	partial pressure gradient	atm
α_{el}	electrode charge transfer coefficient	-
$\gamma(\tau)$	distribution function (analytical)	Ωs
γ_{el}	prefactor of the exchange current density	A/m^2
ε_{el}	electrode porosity	-
ζ	differential RC element	Ω

Symbol	Description	Unit/Value
η_i	polarization overpotential	V
θ_s	sulfur coverage	-
$\nu_{r,k}$	stoichiometric coefficient	-
ξ_{kn}	Knudsen-factor	-
ρ	density of the fuel gas	kg/m^3
σ	conductivity	S/m
τ	characteristic time constant	s
τ_{el}	electrode tortuosity	-
φ	phase angle	degree
χ	differential ohmic resistance	Ω
ψ_{el}	electrode structural parameter	-
ω	angular frequency	rad/s
ω_k	mass fraction	-

H. Acronyms

AFL	anode functional layer
APU	auxiliary power unit
ASC	anode supported cell
ASR	area specific resistance
ATR	autothermal reforming
C/V	current/voltage
CHP	combined heat and power
CNLS	complex nonlinear least squares
CPE	constant phase element
CPO	catalytic partial oxidation
CSM	Colorado School of Mines
DIR	direct internal reforming
DRT	distribution of relaxation times
ECM	equivalent circuit model
EIS	Electrochemical Impedance Spectroscopy
EU	European Union
FEM	finite element method
FIB	focused ion beam
FRA	frequency response analyzer
FZJ	Forschungszentrum Jülich
GDC	gadolinium doped ceria
GFLW	general finite-length Warburg
IIR	indirect internal reforming
ITCP	Institute of Chemical Technology and Polymer Chemistry
IWE	Institut für Werkstoffe der Elektrotechnik

| KIT | Karlsruhe Institute of Technology |

LSC	lanthanum strontium cobaltate
LSCF	lanthanum strontium cobalt ferrite
LSM	lanthanum strontium manganite

MEA	membrane electrode assembly
MFC	mass flow controller
MIEC	mixed ionic electronic conducting

| OCC | open circuit condition |
| OCV | open circuit voltage |

| RC | resistor-capacitor |

SEM	scanning electron microscopy
SOEC	solid oxide electrolysis cell
SOFC	solid oxide fuel cell
SSZ	scandia stabilized zirconia

| TLM | transmission line model |
| TPB | three phase boundary |

| WGS | water-gas shift |

| YSZ | yttria stabilized zirconia |

Bibliography

[1] N. Q. Minh and T. Takahashi. *Science and technology of ceramic fuel cells.* Elsevier, Amsterdam, 1. edition, 1995.

[2] S. C. Singhal and K. Kendall, editors. *High temperature solid oxide fuel cells: fundamentals, design, and applicatons.* Elsevier Advanced Technology, Oxford, UK, 1. edition, 2003.

[3] W. Vielstich, editor. *Handbook of fuel cells: fundamentals, technology and applications.* Wiley, New York, 1. edition, 2003.

[4] M. J. Heneka. *Alterung der Festelektrolyt-Brennstoffzelle unter thermischen und elektrischen Lastwechseln.* PhD thesis, Universität Karlsruhe, 2006.

[5] W. Nernst. Die elektromotorische Wirksamkeit der Jonen. *Zeitschrift für Physikalische Chemie*, 4(2):129–181, 1889.

[6] E. Ivers-Tiffée. Brennstoffzellen und Batterien. Lecture notes, Institut für Werkstoffe der Elektrotechnik (IWE), Karlsruher Institut für Technologie (KIT), 2009.

[7] A. Leonide. *SOFC Modelling and Parameter Identification by Means of Impedance Spectroscopy.* PhD thesis, Karlsruher Institut für Technologie (KIT), Karlsruhe, 2010.

[8] N. Q. Minh. Ceramic Fuel Cells. *Journal of the American Ceramic Society*, 76(3):563–588, 1993.

[9] O. Yamamoto. Solid oxide fuel cells: fundamental aspects and prospects. *Electrochimica Acta*, 45(15-16):2423–2435, 2000.

[10] F. Tietz, H.-P. Buchkremer, and D. Stöver. Components manufacturing for solid oxide fuel cells. *Solid State Ionics*, 152-153(0):373–381, 2002.

[11] S. M. Haile. Fuel cell materials and components. *Acta Materialia*, 51(19):5981–6000, 2003.

[12] N. P. Brandon, S. Skinner, and B. C. H. Steele. Recent advances in materials for fuel cells. *Annual Review of Materials Research*, 33(1):183–213, 2003.

[13] R. M. Ormerod. Solid oxide fuel cells. *Chemical Society Reviews*, 32:17–28, 2003.

[14] A. Weber and E. Ivers-Tiffée. Materials and concepts for solid oxide fuel cells (SOFCs) in stationary and mobile applications. *Journal of Power Sources*, 127(1-2):273–283, 2004.

[15] R. J. Kee, H. Zhu, and D. G. Goodwin. Solid-oxide fuel cells with hydrocarbon fuels. *Proceedings of the Combustion Institute*, 30(2):2379–2404, 2005.

[16] C. Wagner. Über den Mechanismus der elektrischen Stromleitung im Nernststift. *Naturwissenschaften*, 31:265–268, 1943.

[17] C. Wagner. Die Löslichkeit von Wasserdampf in ZrO2-Y2O3-Mischkristallen. *Berichte der Bunsengesellschaft für physikalische Chemie*, 72(7):778–781, 1968.

[18] M. J. Verkerk, B. J. Middelhuis, and A. J. Burggraaf. Effect of grain boundaries on the conductivity of high-purity ZrO2-Y2O3 ceramics. *Solid State Ionics*, 6(2):159–170, 1982.

[19] S. P. S. Badwal. Electrical conductivity of single crystal and polycrystalline yttria-stabilized zirconia. *Journal of Materials Science*, 19:1767–1776, 1984.

[20] S. P. S. Badwal and J. Drennan. Yttria-zirconia: Effect of microstructure on conductivity. *Journal of Materials Science*, 22:3231–3239, 1987.

[21] S. P. S. Badwal. Grain boundary resistivity in zirconia-based materials: effect of sintering temperatures and impurities. *Solid State Ionics*, 76(1-2):67–80, 1995.

[22] U. B. Pal and S. C. Singhal. Electrochemical Vapor Deposition of Yttria-Stabilized Zirconia Films. *Journal of The Electrochemical Society*, 137(9):2937–2941, 1990.

[23] S. de Souza, S. J. Visco, and L. C. De Jonghe. Thin-film solid oxide fuel cell with high performance at low-temperature. *Solid State Ionics*, 98(1-2):57–61, 1997.

[24] P. Bohac and L. J. Gauckler. Chemical spray deposition of YSZ and GCO solid electrolyte films. *Solid State Ionics*, 119(1-4):317–321, 1999.

[25] K. Nomura, Y. Mizutani, M. Kawai, Y. Nakamura, and O. Yamamoto. Aging and Raman scattering study of scandia and yttria doped zirconia. *Solid State Ionics*, 132(3-4):235–239, 2000.

[26] K. Sasaki and J. Maier. Re-analysis of defect equilibria and transport parameters in Y2O3-stabilized ZrO2 using EPR and optical relaxation. *Solid State Ionics*, 134(3-4):303–321, 2000.

[27] E. Wanzenberg, F. Tietz, D. Kek, P. Panjan, and D. Stöver. Influence of electrode contacts on conductivity measurements of thin YSZ electrolyte films and the impact on solid oxide fuel cells. *Solid State Ionics*, 164(3-4):121–129, 2003.

[28] B. Butz, P. Kruse, H. Störmer, D. Gerthsen, A. Müller, A. Weber, and E. Ivers-Tiffée. Correlation between microstructure and degradation in conductivity for cubic Y2O3-doped ZrO2. *Solid State Ionics*, 177(37-38):3275–3284, 2006.

[29] M. Mori, T. Yamamoto, H. Itoh, H. Inaba, and H. Tagawa. Thermal Expansion of Nickel-Zirconia Anodes in Solid Oxide Fuel Cells during Fabrication and Operation. *Journal of The Electrochemical Society*, 145(4):1374–1381, 1998.

[30] P. Costamagna, P. Costa, and V. Antonucci. Micro-modelling of solid oxide fuel cell electrodes. *Electrochimica Acta*, 43(3-4):375–394, 1998.

[31] W. Z. Zhu and S. C. Deevi. A review on the status of anode materials for solid oxide fuel cells. *Materials Science and Engineering: A*, 362(1-2):228–239, 2003.

[32] A. Atkinson, S. A. Barnett, R. J. Gorte, J. T. S. Irvine, A. J. McEvoy, M. Mogensen, S. C. Singhal, and J. Vohs. Advanced anodes for high-temperature fuel cells. *Nature Materials*, Jan; 3:17–27, 2004.

[33] S. P. Jiang and S. H. Chan. A review of anode materials development in solid oxide fuel cells. *Journal of Materials Science*, 39:4405–4439, 2004.

[34] C. Sun and U. Stimming. Recent anode advances in solid oxide fuel cells. *Journal of Power Sources*, 171(2):247–260, 2007.

[35] F. Tietz, F. J. Dias, D. Simwonis, and D. Stöver. Evaluation of commercial nickel oxide powders for components in solid oxide fuel cells. *Journal of the European Ceramic Society*, 20(8):1023–1034, 2000.

[36] D. Fouquet, A. C. Müller, A. Weber, and E. Ivers-Tiffée. Kinetics of oxidation and reduction of Ni/YSZ cermets. *Ionics*, 9:103–108, 2003.

[37] D. Sarantaridis and A. Atkinson. Redox Cycling of Ni-Based Solid Oxide Fuel Cell Anodes: A Review. *Fuel Cells*, 7(3):246–258, 2007.

[38] Q. Fu, F. Tietz, D. Sebold, Sh. Tao, and J. T. S. Irvine. An Efficient Ceramic-based Anode for Solid Oxide Fuel Cells. *Journal of Power Sources*, 171(2):663–669, 2007.

[39] Q. Ma, F. Tietz, A. Leonide, and E. Ivers-Tiffée. Anode-supported planar sofc with high performance and redox stability. *Electrochemistry Communications*, 12(10):1326–1328, 2010.

[40] Q. Ma and F. Tietz. Comparison of Y and La-substituted SrTiO3 as the anode materials for SOFCs. *Solid State Ionics*, 225(0):108–112, 2012.

[41] M. C. Tucker. Progress in metal-supported solid oxide fuel cells: A review. *Journal of Power Sources*, 195(15):4570–4582, 2010.

[42] I. Villarreal, C. Jacobson, A. Leming, Y. Matus, S. Visco, and L. C. De Jonghe. Metal-Supported Solid Oxide Fuel Cells. *Electrochemical and Solid-State Letters*, 6(9):A178–A179, 2003.

[43] R. Hui, D. Yang, Z. Wang, S. Yick, C. Decès-Petit, W. Qu, A. Tuck, R. Maric, and D. Ghosh. Metal-supported solid oxide fuel cell operated at 400 - 600°C. *Journal of Power Sources*, 167(2):336–339, 2007.

[44] G. Schiller, A. Ansar, M. Lang, and O. Patz. High temperature water electrolysis using metal supported solid oxide electrolyser cells (SOEC). *Journal of Applied Electrochemistry*, 39:293–301, 2009.

[45] P. Blennow, J. Hjelm, T. Klemensø, S. Ramousse, A. Kromp, A. Leonide, and A. Weber. Manufacturing and characterization of metal-supported solid oxide fuel cells. *Journal of Power Sources*, 196(17):7117–7125, 2011.

[46] A. Kromp, P. Blennow, J. Nielsen, T. Klemensø, and A. Weber. Break-down of Losses in High Performing Metal-Supported Solid Oxide Fuel Cells. *Fuel Cells*, 2013. Published online, DOI: 10.1002/fuce.201200165.

[47] F. Tietz, A. Mai, and D. Stöver. From powder properties to fuel cell performance - A holistic approach for SOFC cathode development. *Solid State Ionics*, 179(27-32):1509–1515, 2008.

[48] B. Rüger. *Mikrostrukturmodellierung von Elektroden für die Festelektrolytbrenn-stoffzelle*. PhD thesis, Universität Karlsruhe, 2009.

[49] J. Richter, P. Holtappels, T. Graule, T. Nakamura, and L. J. Gauckler. Materials design for perovskite SOFC cathodes. *Monatshefte für Chemie*, 140:985–999, 2012.

[50] C. Endler-Schuck. *Alterungsverhalten mischleitender LSCF Kathoden für Hochtemperatur-Festoxid-Brennstoffzellen (SOFCs)*. PhD thesis, Karlsruher Institut für Technologie (KIT), 2012.

[51] O. Yamamoto, Y. Takeda, R. Kanno, and M. Noda. Perovskite-type oxides as oxygen electrodes for high temperature oxide fuel cells. *Solid State Ionics*, 22(2-3):241–246, 1987.

[52] S. Uhlenbruck, N. Jordan, D. Sebold, H. P. Buchkremer, V. A. C. Haanappel, and D. Stöver. Thin film coating technologies of (Ce,Gd)O2-d interlayers for application in ceramic high-temperature fuel cells. *Thin Solid Films*, 515(7-8):4053–4060, 2007.

[53] A. Mai, V. A. C. Haanappel, F. Tietz, and D. Stöver. Ferrite-based Perovskites as Cathode Materials for Anode-supported Solid Oxide Fuel Cells: Part II. Influence of the CGO Interlayer. *Solid State Ionics*, 177(19-25):2103–2107, 2006.

[54] A. Mai, V. A. C. Haanappel, S. Uhlenbruck, F. Tietz, and D. Stöver. Ferrite-based Perovskites as Cathode Materials for Anode-supported Solid Oxide Fuel Cells: Part I. Variation of Composition. *Solid State Ionics*, 176(15-16):1341–1350, 2005.

[55] F. Tietz, V. A. C. Haanappel, A. Mai, J. Mertens, and D. Stöver. Performance of LSCF cathodes in cell tests. *Journal of Power Sources*, 156(1):20–22, 2006.

[56] A. Mai, M. Becker, W. Assenmacher, F. Tietz, D. Hathiramani, E. Ivers-Tiffée, D. Stöver, and W. Mader. Time-dependent performance of mixed-conducting SOFC cathodes. *Solid State Ionics*, 177(19-25):1965–1968, 2006.

[57] C. Endler-Schuck, A. Leonide, A. Weber, F. Tietz, and E. Ivers-Tiffée. Time-Dependent Electrode Performance Changes in Intermediate Temperature Solid Oxide Fuel Cells. *Journal of The Electrochemical Society*, 157(2):B292–B298, 2010.

[58] C. Endler-Schuck, A. Leonide, A. Weber, S. Uhlenbruck, F. Tietz, and E. Ivers-Tiffée. Performance analysis of mixed ionic-electronic conducting cathodes in anode supported cells. *Journal of Power Sources*, 196(17):7257–7262, 2011.

[59] A. Leonide, Y. Apel, and E. Ivers-Tiffée. SOFC Modeling and Parameter Identification by Means of Impedance Spectroscopy. *ECS Transactions*, 19(20):81–109, 2009.

[60] A. Leonide, S. Hansmann, and E. Ivers-Tiffée. A 0-Dimensional Stationary Model for Anode-Supported Solid Oxide Fuel Cells. *ECS Transactions*, 28(11):341–346, 2010.

[61] E. Ivers-Tiffée, A. Weber, and H. Schichlein. Electrochemical impedance spectroscopy. In W. Vielstich, H. A. Gasteiger, and A. Lamm, editors, *Handbook of Fuel Cells — Fundamentals, Technology and Applications, Vol. 2*, pages 220–235. John Wiley & Sons, Ltd, 2003.

[62] J. R. Macdonald, editor. *Impedance spectroscopy: emphasizing solid materials and systems*. Wiley-Interscience, New York, 1987.

[63] M. E. Orazem and B. Tribollet. *Electrochemical Impedance Spectroscopy.* The Electrochemical Society series. Wiley, Hoboken, 2008.

[64] V. F. Lvovich. *Impedance Spectroscopy: Applications to Electrochemical and Dielectric Phenomena.* John Wiley & Sons, Inc., New York, 1. edition, 2012.

[65] H. Schichlein. *Experimentelle Modellbildung für die Hochtemperatur-Brennstoffzelle SOFC.* PhD thesis, Universität Karlsruhe, 2003.

[66] M. Kluwe. Systemdynamik und Regelungstechnik. Lecture notes, Institut für Regelungs- und Steuerungssysteme (IRS), Karlsruher Institut für Technologie (KIT), 2009.

[67] A. C. Müller. *Mehrschicht-Anode für die Hochtemperatur-Brennstoffzelle (SOFC).* PhD thesis, Universität Karlsruhe, 2005.

[68] R. Steinberger-Wilckens, L. Blum, H.-P. Buchkremer, B. De Haart, J. Malzbender, and M. Pap. Recent Results in Solid Oxide Fuel Cell Development at Forschungszentrum Jülich. *ECS Transactions*, 35(1):53–60, 2011.

[69] D. Klotz. *Characterization and Modeling of Electrochemical Energy Conversion Systems by Impedance Techniques.* PhD thesis, Karlsruher Institut für Technologie (KIT), 2012.

[70] H. W. Engl, M. Hanke-Bourgeois, and A. Neubauer. *Regularization of inverse problems*, volume 375 of *Mathematics and its applications.* Kluwer Academic Publishing, Dordrecht, 2000.

[71] H. W. Engl. Inverse problems and their regularization. In R. E. Burkard, A. Jameson, G. Strang, P. Deuflhard, J.-L. Lions, V. Capasso, J. Periaux, and H. W. Engl, editors, *Computational Mathematics Driven by Industrial Problems*, volume 1739 of *Lecture Notes in Mathematics*, pages 127–150. Springer, Heidelberg, 2000.

[72] A. N. Tikhonov and V. J. Arsenin. *Solutions of ill-posed problems.* Scripta series in mathematics. A Halsted Press book. Winston, Washington, D.C., 1977.

[73] A. N. Tikhonov, editor. *Numerical Methods for the Solution of Ill-posed Problems.* Mathematics and its Applications. Kluwer Academic, Dordrecht, 1995.

[74] J. Weese. A Reliable and Fast Method for the Solution of Fredholm Integral Equations of the First Kind based on Tikhonov Regularization. *Computer Physics Communications*, 69(1):99–111, 1992.

[75] H. Schichlein, A.C. Müller, M. Voigts, A. Krügel, and E. Ivers-Tiffée. Deconvolution of Electrochemical Iimpedance Spectra for the Identification of Electrode Reaction Mechanisms in Solid Oxide Fuel Cells. *Journal of Applied Electrochemistry*, 32:875–882, 2002.

[76] C. H. Hamann and W. Vielstich. *Elektrochemie II: Elektrodenprozesse, angewandte Elektrochemie.* Taschentext 42. Verlag Chemie, Weinheim, 1981.

[77] C. H. Hamann, A. Hamnett, and W. Vielstich. *Electrochemistry.* Wiley-VCH, Weinheim, 2. edition, 2007.

[78] A. J. Bard and L. R. Faulkner. *Electrochemical Methods: Fundamentals and Applications.* Wiley, New York, 2. edition, 2001.

[79] H. Wendt and G. Kreysa. *Electrochemical engineering - science and technology in chemical and other industries.* Springer, Berlin, 1999.

[80] J. S. Newman and C. W. Tobias. Theoretical Analysis of Current Distribution in Porous Electrodes. *Journal of The Electrochemical Society*, 109(12):1183–1191, 1962.

[81] J. Bisquert, G. Garcia-Belmonte, F. Fabregat-Santiago, and A. Compte. Anomalous Transport Effects in the Impedance of Porous Film Electrodes. *Electrochemistry Communications*, 1(9):429–435, 1999.

[82] J. Bisquert. Influence of the Boundaries in the Impedance of Porous Film Electrodes. *Physical Chemistry Chemical Physics*, 2:4185–4192, 2000.

[83] S. B. Adler, J. A. Lane, and B. C. H. Steele. Electrode Kinetics of Porous Mixed-Conducting Oxygen Electrodes. *Journal of The Electrochemical Society*, 143(11):3554–3564, 1996.

[84] B. A. Boukamp and H. J. M. Bouwmeester. Interpretation of the Gerischer Impedance in Solid State Ionics. *Solid State Ionics*, 157(1-4):29–33, 2003.

[85] A. Leonide, B. Rüger, A. Weber, W. A. Meulenberg, and E. Ivers-Tiffée. Impedance Study of Alternative $(La,Sr)FeO_{3-\delta}$ and $(La,Sr)(Co,Fe)O_{3-\delta}$ MIEC Cathode Compositions. *Journal of The Electrochemical Society*, 157(2):B234–B239, 2010.

[86] A. Leonide, V. Sonn, A. Weber, and E. Ivers-Tiffée. Evaluation and Modeling of the Cell Resistance in Anode-Supported Solid Oxide Fuel Cells. *Journal of The Electrochemical Society*, 155(1):B36–B41, 2008.

[87] V. Sonn, A. Leonide, and E. Ivers-Tiffée. Combined Deconvolution and CNLS Fitting Approach Applied on the Impedance Response of Technical Ni/8YSZ Cermet Electrodes. *Journal of The Electrochemical Society*, 155(7):B675–B679, 2008.

[88] A. Leonide, S. Ngo Dinh, A. Weber, and E. Ivers-Tiffée. Performance Limiting Factors in Anode Supported SOFC. *Proceedings of the 8th European Solid Oxide Fuel Cell Forum*, page A0501, 2008.

[89] E. L. Cussler. *Diffusion: mass transfer in fluid systems.* Cambridge University Press, Cambridge, 3. ed. edition, 2009.

[90] J. W. Kim, A. V. Virkar, K. Z. Fung, K. Mehta, and S. C. Singhal. Polarization Effects in Intermediate Temperature, Anode-Supported Solid Oxide Fuel Cells. *Journal of The Electrochemical Society*, 146(1):69–78, 1999.

[91] S. Primdahl and M. Mogensen. Gas diffusion impedance in characterization of solid oxide fuel cell anodes. *Journal of the Electrochemical Society*, 146(8):2827–2833, 1999.

[92] S. Primdahl and M. Mogensen. Gas Conversion Impedance: A Test Geometry Effect in Characterization of Solid Oxide Fuel Cell Anodes. *Journal of the Electrochemical Society*, 145(7):2431–2438, 1998.

[93] J. L. Dawson and D. G. John. Diffusion impedance - An extended general analysis. *Journal of Electroanalytical Chemistry and Interfacial Electrochemistry*, 110(1-3):37–47, 1980.

[94] B. C. H. Steele. Fuel-cell technology: Running on natural gas. *Nature*, 400:619–620, 1999.

[95] J. Van Herle, N. Autissier, D. Favrat, D. Larrain, Z. Wuillemin, and M. Molinelli. Modeling and Experimental Validation of Solid Oxide Fuel Cell Materials and Stacks. *Journal of the European Ceramic Society*, 25(12):2627–2632, 2005.

[96] Z. Zhan, J. Liu, and S. A. Barnett. Operation of anode-supported solid oxide fuel cells on propane-air fuel mixtures. *Applied Catalysis A: General*, 262(2):255–259, 2004.

[97] A. Lindermeir, S. Kah, S. Kavurucu, and M. Mühlner. On-board diesel fuel processing for an SOFC-APU — Technical challenges for catalysis and reactor design. *Applied Catalysis B: Environmental*, 70(1-4):488–497, 2007.

[98] P. K. Cheekatamarla, C. M. Finnerty, C. R. Robinson, S. M. Andrews, J. A. Brodie, Y. Lu, and P. G. DeWald. Design, integration and demonstration of a 50 kW JP8/kerosene fueled portable SOFC power generator. *Journal of Power Sources*, 193(2):797–803, 2009.

[99] M. Santin, A. Traverso, and L. Magistri. Liquid fuel utilization in SOFC hybrid systems. *Applied Energy*, 86(10):2204–2212, 2009.

[100] O. A. Marina, C. A. Coyle, E. C. Thomsen, D. J. Edwards, G. W. Coffey, and L. R. Pederson. Degradation mechanisms of SOFC anodes in coal gas containing phosphorus. *Solid State Ionics*, 181(8-10):430–440, 2010.

[101] A. L. Dicks. Hydrogen generation from natural gas for the fuel cell systems of tomorrow. *Journal of Power Sources*, 61(1-2):113–124, 1996.

[102] S. H. Clarke, A. L. Dicks, K. Pointon, T. A. Smith, and A. Swann. Catalytic aspects of the steam reforming of hydrocarbons in internal reforming fuel cells. *Catalysis Today*, 38(4):411–423, 1997.

[103] A. L. Dicks. Advances in catalysts for internal reforming in high temperature fuel cells. *Journal of Power Sources*, 71(1-2):111–122, 1998.

[104] J. R. Rostrup-Nielsen. Conversion of hydrocarbons and alcohols for fuel cells. *Physical Chemistry Chemical Physics*, 3:283–288, 2001.

[105] M. Mogensen and K. Kammer. Conversion of Hydrocarbons in Solid Oxide Fuel Cells. *Annual Review of Materials Research*, 33(1):321–331, 2003.

[106] J. P. Van Hook. Methane-Steam Reforming. *Catalysis Reviews*, 21(1):1–51, 1980.

[107] J. R. Rostrup-Nielsen. Catalytic Steam Reforming. In J. R. Anderson and M. Boudart, editors, *Catalysis, Science and Technology*, pages 1–117. Springer, Berlin, 1983.

[108] V. Twigg. *Catalyst handbook*. Wolfe, 1989.

[109] F. Fischer and H. Tropsch. Conversion of methane into hydrogen and carbon monoxide. *Brennst. Chem.*, 3(9):39–46, 1928. cited By (since 1996) 60.

[110] S. S. Bharadwaj and L. D. Schmidt. Catalytic partial oxidation of natural gas to syngas. *Fuel Processing Technology*, 42(2-3):109–127, 1995.

[111] B. C. Enger, R. Løldeng, and A. Holmen. A review of catalytic partial oxidation of methane to synthesis gas with emphasis on reaction mechanisms over transition metal catalysts. *Applied Catalysis A: General*, 346(1-2):1–27, 2008.

[112] A. Weber, B. Sauer, A. C. Müller, D. Herbstritt, and E. Ivers-Tiffée. Oxidation of H2, CO and methane in SOFCs with Ni/YSZ-cermet anodes. *Solid State Ionics*, 152-153(0):543–550, 2002.

[113] J. Liu and S. A. Barnett. Operation of anode-supported solid oxide fuel cells on methane and natural gas. *Solid State Ionics*, 158(1-2):11–16, 2003.

[114] Y. Lin, Z. Zhan, J. Liu, and S. A. Barnett. Direct operation of solid oxide fuel cells with methane fuel. *Solid State Ionics*, 176(23-24):1827–1835, 2005.

[115] E. P. Murray, T. Tsai, and S. A. Barnett. A direct-methane fuel cell with a ceria-based anode. *Nature*, 400:649–651, 1999.

[116] S. McIntosh and R. J. Gorte. Direct Hydrocarbon Solid Oxide Fuel Cells. *Chemical Reviews*, 104(10):4845–4866, 2004. PMID: 15669170.

[117] C. M. Finnerty, N. J. Coe, R. H. Cunningham, and R. M. Ormerod. Carbon formation on and deactivation of nickel-based/zirconia anodes in solid oxide fuel cells running on methane. *Catalysis Today*, 46(2-3):137–145, 1998.

[118] H. Timmermann, W. Sawady, D. Campbell, A. Weber, R. Reimert, and E. Ivers-Tiffée. Coke Formation and Degradation in SOFC Operation with a Model Reformate from Liquid Hydrocarbons. *Journal of The Electrochemical Society*, 155(4):B356–B359, 2008.

[119] D. Mogensen, J.-D. Grunwaldt, P. V. Hendriksen, K. Dam-Johansen, and J. U. Nielsen. Internal steam reforming in solid oxide fuel cells: Status and opportunities of kinetic studies and their impact on modelling. *Journal of Power Sources*, 196(1):25–38, 2011.

[120] W. Lehnert, J. Meusinger, and F. Thom. Modelling of gas transport phenomena in SOFC anodes. *Journal of Power Sources*, 87(1-2):57–63, 2000.

[121] P. Aguiar, C. S. Adjiman, and N. P. Brandon. Anode-supported intermediate temperature direct internal reforming solid oxide fuel cell. I: model-based steady-state performance. *Journal of Power Sources*, 138(1-2):120–136, 2004.

[122] E. S. Hecht, G. K. Gupta, H. Zhu, A. M. Dean, R. J. Kee, L. Maier, and O. Deutschmann. Methane reforming kinetics within a Ni–YSZ SOFC anode support. *Applied Catalysis A: General*, 295(1):40–51, 2005.

[123] V. M. Janardhanan and O. Deutschmann. CFD analysis of a solid oxide fuel cell with internal reforming: Coupled interactions of transport, heterogeneous catalysis and electrochemical processes. *Journal of Power Sources*, 162(2):1192–1202, 2006.

[124] H. Timmermann, W. Sawady, R. Reimert, and E. Ivers-Tiffée. Kinetics of (reversible) internal reforming of methane in solid oxide fuel cells under stationary and APU conditions. *Journal of Power Sources*, 195(1):214–222, 2010.

[125] H. Timmermann. *Untersuchungen zum Einsatz von Reformat aus flüssigen Kohlenwasserstoffen in der Hochtemperaturbrennstoffzelle SOFC*. PhD thesis, Karlsruher Institut für Technologie (KIT), 2010.

[126] A. Kromp, A. Leonide, H. Timmermann, A. Weber, and E. Ivers-Tiffée. Internal Reforming Kinetics in SOFC-Anodes. *ECS Transactions*, 28(11):205–215, 2010.

[127] Y. Jiang and A. V. Virkar. Fuel Composition and Diluent Effect on Gas Transport and Performance of Anode-Supported SOFCs. *Journal of The Electrochemical Society*, 150(7):A942–A951, 2003.

[128] H. Zhu and R. J. Kee. Modeling electrochemical impedance spectra in sofc button cells with internal methane reforming. *Journal of The Electrochemical Society*, 153(9):A1765–A1772, 2006.

[129] S. Zha, Z. Cheng, and M. Liu. Sulfur Poisoning and Regeneration of Ni-Based Anodes in Solid Oxide Fuel Cells. *Journal of The Electrochemical Society*, 154(2):B201–B206, 2007.

[130] M. Gong, X. Liu, J. Trembly, and C. Johnson. Sulfur-tolerant anode materials for solid oxide fuel cell application. *Journal of Power Sources*, 168(2):289–298, 2007.

[131] Z. Cheng, J. H. Wang, Y. M. Choi, L. Yang, M. C. Lin, and M. Liu. From Ni-YSZ to sulfur-tolerant anode materials for SOFCs: electrochemical behavior, in situ characterization, modeling, and future perspectives. *Energy Environmental Science*, 4:4380–4409, 2011.

[132] K. Sasaki. Thermochemical Stability of Sulfur Compounds in Fuel Cell Gases Related to Fuel Impurity Poisoning. *Journal of Fuel Cell Science and Technology*, 5(3):031212–1 – 031212–8, 2008.

[133] E. Brightman, D. G. Ivey, D. J. L. Brett, and N. P. Brandon. The effect of current density on H_2S-poisoning of nickel-based solid oxide fuel cell anodes. *Journal of Power Sources*, 196(17):7182–7187, 2011.

[134] C. H. Bartholomew, P. K. Agrawal, and J. R. Katzer. Sulfur Poisoning of Metals. volume 31 of *Advances in Catalysis*, pages 135–242. Academic Press, 1982.

[135] D. R. Huntley. The decomposition of H_2S on Ni(110). *Surface Science*, 240(1-3):13–23, 1990.

[136] C. H. Bartholomew. Mechanisms of catalyst deactivation. *Applied Catalysis A: General*, 212(1-2):17–60, 2001.

[137] K. Sasaki, K. Susuki, A. Iyoshi, M. Uchimura, N. Imamura, H. Kusaba, Y. Teraoka, H. Fuchino, K. Tsujimoto, Y. Uchida, and N. Jingo. H_2S Poisoning of Solid Oxide Fuel Cells. *Journal of The Electrochemical Society*, 153(11):A2023–A2029, 2006.

[138] T. S. Li, W. G. Wang, T. Chen, H. Miao, and C. Xu. Hydrogen sulfide poisoning in solid oxide fuel cells under accelerated testing conditions. *Journal of Power Sources*, 195(20):7025–7032, 2010.

[139] J. F. B. Rasmussen and A. Hagen. The effect of H_2S on the performance of Ni-YSZ anodes in solid oxide fuel cells. *Journal of Power Sources*, 191(2):534–541, 2009.

[140] Y. Matsuzaki and I. Yasuda. The poisoning effect of sulfur-containing impurity gas on a SOFC anode: Part I. Dependence on temperature, time, and impurity concentration. *Solid State Ionics*, 132(3-4):261–269, 2000.

[141] P. Lohsoontorn, D. J. L. Brett, and N. P. Brandon. The effect of fuel composition and temperature on the interaction of H2S with nickel-ceria anodes for Solid Oxide Fuel Cells. *Journal of Power Sources*, 183(1):232–239, 2008.

[142] A. Hagen, J. F. B. Rasmussen, and K. Thydén. Durability of solid oxide fuel cells using sulfur containing fuels. *Journal of Power Sources*, 196(17):7271–7276, 2011.

[143] H. P. He, A. Wood, D. Steedman, and M. Tilleman. Sulphur tolerant shift reaction catalysts for nickel-based SOFC anode. *Solid State Ionics*, 179(27-32):1478–1482, 2008.

[144] L. Blum, H.-P. Buchkremer, S. M. Gross, B. De Haart, J. W. Quadakkers, U. Reisgen, R. Steinberger-Wilckens, R. W. Steinbrech, and F. Tietz. Overview of the Development of Solid Oxide Fuel Cells at Forschungszentrum Jülich. *Proceedings - Electrochemical Society*, PV 2005-07:39–47, 2005.

[145] D. Stöver, H. P. Buchkremer, and J. P. P. Huijsmans. MEA/cell preparation methods: Europe/USA. In W. Vielstich, H. A. Gasteiger, and A. Lamm, editors, *Handbook of Fuel Cells — Fuel Cell Technology and Applications Vol. 4*, pages 1015–1029. John Wiley & Sons, Ltd, 2003.

[146] D. Stöver, H. P. Buchkremer, F.Tietz, and N. H. Menzler. Trends in processing of SOFC components. *Proceedings of the 5th European Solid Oxide Fuel Cell Forum*, pages 1–9, 2002.

[147] R. Mücke, N. H. Menzler, H. P. Buchkremer, and D. Stöver. Cofiring of Thin Zirconia Films During SOFC Manufacturing. *Journal of the American Ceramic Society*, 92:S95–S102, 2009.

[148] N. H. Menzler, W. Schafbauer, F. Han, O. Büchler, R. Mücke, H. P. Buchkremer, and D. Stöver. Development of High Power Density Solid Oxide Fuel Cells (SOFCs) for Long-Term Operation. *Materials Science Forum*, 654-656:2875–2878, 2010.

[149] A. Weber. *Entwicklung von Kathodenstrukturen für die Hochtemperatur-Brennstoffzelle SOFC*. PhD thesis, Universität Karlsruhe, 2003.

[150] M. Becker, A. Mai, E. Ivers-Tiffée, and F. Tietz. Long-Term Measurements of Anode-supported Solid Oxide Fuel Cells with LSCF Cathode under various Operating Conditions. *Proceedings of the Electrochemical Society*, 7:514–523, 2005.

[151] M. Kornely. *Elektrische Charakterisierung und Modellierung von metallischen Interkonnektoren (MIC) des SOFC-Stacks*. PhD thesis, Karlsruher Institut für Technologie (KIT), 2012.

[152] H. Yokokawa, S. Yamauchi, and T. Matsumoto. Thermodynamic database MALT for Windows with gem and CHD. *Calphad*, 26(2):155–166, 2002.

[153] B. A. Boukamp. A Linear Kronig-Kramers Transform Test for Immittance Data Validation. *Journal of The Electrochemical Society*, 142(6):1885–1894, 1995.

[154] B. A. Boukamp. Electrochemical Impedance Spectroscopy in Solid State Ionics: Recent Advances. *Solid State Ionics*, 169(1-4):65–73, 2004.

[155] A. Kromp, A. Leonide, A. Weber, and E. Ivers-Tiffée. Electrochemical Analysis of Reformate-Fuelled Anode Supported SOFC. *Journal of The Electrochemical Society*, 158(8):B980–B986, 2011.

[156] S. B. Adler. Mechanism and Kinetics of Oxygen Reduction on Porous $La_{1-x}Sr_xCoO_{3-d}$ Electrodes. *Solid State Ionics*, 111(1-2):125–134, 1998.

[157] H. Geisler. *Transiente Simulation der Stofftransportvorgänge im Anodensubstrat der SOFC*. Diploma thesis, Karlsruher Institut für Technologie (KIT), 2011.

[158] A. Kromp, H. Geisler, A. Weber, and E. Ivers-Tiffée. Electrochemical Impedance Modeling of Reformate- Fuelled Anode-Supported SOFC. *Proceedings of the 10th European Solid Oxide Fuel Cell Forum*, page B1092, 2012.

[159] W. G. Bessler, S. Gewies, and M. Vogler. A New Framework for Physically based Modeling of Solid Oxide Fuel Cells. *Electrochimica Acta*, 53(4):1782–1800, 2007.

[160] H. Zhu, A. Kromp, A. Leonide, E. Ivers-Tiffée, O. Deutschmann, and R. J. Kee. A Model-Based Interpretation of the Influence of Anode Surface Chemistry on Solid Oxide Fuel Cell Electrochemical Impedance Spectra. *Journal of The Electrochemical Society*, 159(7):F255–F266, 2012.

[161] C. F. Curtiss and R. B. Bird. Multicomponent Diffusion. *Industrial & Engineering Chemistry Research*, 38(7):2515–2522, 1999.

[162] R. B. Bird, W. E. Stewart, and E. N. Lightfoot. *Transport phenomena*. Wiley, New York, 1960.

[163] COMSOL Multiphysics. *Version 4.1*. COMSOL AB, Stockholm, Sweden, 2010.

[164] S. Chapman, T. G. Cowling, and C. Cercignani. *The Mathematical Theory of Nonuniform Gases: An Account of the Kinetic Theory of Viscosity, Thermal Conduction and Diffusion in Gases*. Cambridge Mathematical Library. Cambridge University Press, 1991.

[165] S. Dierickx. *Schwefelvergiftung im Reformatbetrieb der Hochtemperaturbrennstoffzelle SOFC*. Bachelor thesis, Karlsruher Institut für Technologie (KIT), 2011.

[166] A. Kromp, S. Dierickx, A. Leonide, A. Weber, and E. Ivers-Tiffée. Electrochemical Analysis of Sulfur-Poisoning in Anode Supported SOFCs Fuelled with a Model Reformate. *Journal of The Electrochemical Society*, 159(5):B597–B601, 2012.

[167] J. B. Hansen. Correlating Sulfur Poisoning of SOFC Nickel Anodes by a Temkin Isotherm. *Electrochemical and Solid-State Letters*, 11(10):B178–B180, 2008.

[168] I. Alstrup, J. R. Rostrup-Nielsen, and S. Røen. High temperature hydrogen sulfide chemisorption on nickel catalysts. *Applied Catalysis*, 1(5):303–314, 1981.

[169] J. R. Rostrup-Nielsen, J. B. Hansen, S. Helveg, N. Christiansen, and A.-K. Jannasch. Sites for catalysis and electrochemistry in solid oxide fuel cell (SOFC) anode. *Applied Physics A: Materials Science & Processing*, 85:427–430, 2006.

[170] E. A. Mason and A. P. Malinauskas. *Gas transport in porous media: the dusty-gas model*. Chemical engineering monographs ; 17. Elsevier, Amsterdam, 1983.

[171] O. Deutschmann, editor. *Modeling and simulation of heterogeneous catalytic reactions: From the molecular process to the technical system*. Wiley-VCH Verlag GmbH & Co. KGaA, Weinheim, 1. edition, 2011.

[172] V. Janardhanan. *A detailed approach to model transport, heterogeneous chemistry, and electrochemistry in solid-oxide fuel cells*. PhD thesis, Universität Karlsruhe (TH), Karlsruhe, 2007.

[173] A. Weber, S. Dierickx, A. Kromp, and E. Ivers-Tiffée. Sulphur Poisoning of Anode-Supported SOFCs under Reformate Operation. *Fuel Cells*, 2013. Published online, DOI: 10.1002/fuce.201200180.

Werkstoffwissenschaft @ Elektrotechnik /
Universität Karlsruhe, Institut für Werkstoffe der Elektrotechnik

Die Bände sind im Verlagshaus Mainz (Aachen) erschienen.

Band 1   Helge Schichlein
          Experimentelle Modellbildung für die Hochtemperatur-Brennstoff-
          zelle SOFC. 2003
          ISBN 3-86130-229-2

Band 2   Dirk Herbstritt
          Entwicklung und Optimierung einer leistungsfähigen Kathoden-
          struktur für die Hochtemperatur-Brennstoffzelle SOFC. 2003
          ISBN 3-86130-230-6

Band 3   Frédéric Zimmermann
          Steuerbare Mikrowellendielektrika aus ferroelektrischen
          Dickschichten. 2003
          ISBN 3-86130-231-4

Band 4   Barbara Hippauf
          Kinetik von selbsttragenden, offenporösen Sauerstoffsensoren
          auf der Basis von $Sr(Ti,Fe)O_3$. 2005
          ISBN 3-86130-232-2

Band 5   Daniel Fouquet
          Einsatz von Kohlenwasserstoffen in der Hochtemperatur-Brennstoff-
          zelle SOFC. 2005
          ISBN 3-86130-233-0

Band 6   Volker Fischer
          Nanoskalige Nioboxidschichten für den Einsatz in hochkapazitiven
          Niob-Elektrolytkondensatoren. 2005
          ISBN 3-86130-234-9

Band 7   Thomas Schneider
          Strontiumtitanferrit-Abgassensoren.
          Stabilitätsgrenzen / Betriebsfelder. 2005
          ISBN 3-86130-235-7

Band 8   Markus J. Heneka
          Alterung der Festelektrolyt-Brennstoffzelle unter thermischen
          und elektrischen Lastwechseln. 2006
          ISBN 3-86130-236-5

Band 9   Thilo Hilpert
          Elektrische Charakterisierung von Wärmedämmschichten mittels
          Impedanzspektroskopie. 2007
          ISBN 3-86130-237-3

Band 10   Michael Becker
           Parameterstudie zur Langzeitbeständigkeit von Hochtemperatur-
           brennstoffzellen (SOFC). 2007
           ISBN 3-86130-239-X

Band 11   Jin Xu
           Nonlinear Dielectric Thin Films for Tunable Microwave Applications.
           2007
           ISBN 3-86130-238-1

Band 12   Patrick König
           **Modellgestützte Analyse und Simulation von stationären Brennstoff-
           zellensystemen.** 2007
           ISBN 3-86130-241-1

Band 13   Steffen Eccarius
           Approaches to Passive Operation of a Direct Methanol Fuel Cell. 2007
           ISBN 3-86130-242-X

Fortführung als

Schriften des Instituts für Werkstoffe der Elektrotechnik, Karlsruher Institut für Technologie (1868-1603)

bei KIT Scientific Publishing

Die Bände sind unter www.ksp.kit.edu als PDF frei verfügbar oder
als Druckausgabe bestellbar.

Band 14   Stefan F. Wagner
           **Untersuchungen zur Kinetik des Sauerstoffaustauschs an
           modifizierten Perowskitgrenzflächen.** 2009
           ISBN 978-3-86644-362-4

Band 15   Christoph Peters
           **Grain-Size Effects in Nanoscaled Electrolyte and Cathode Thin Films
           for Solid Oxide Fuel Cells (SOFC).** 2009
           ISBN 978-3-86644-336-5

Band 16   Bernd Rüger
           **Mikrostrukturmodellierung von Elektroden für die Festelektro-
           lytbrennstoffzelle.** 2009
           ISBN 978-3-86644-409-6

Band 17   Henrik Timmermann
           **Untersuchungen zum Einsatz von Reformat aus flüssigen Kohlen-
           wasserstoffen in der Hochtemperaturbrennstoffzelle SOFC.** 2010
           ISBN 978-3-86644-478-2

Band 18   André Leonide
           SOFC Modelling and Parameter Identification by Means of Impedance
           Spectroscopy. 2010
           ISBN 978-3-86644-538-3

Band 19   Cornelia Endler-Schuck
           Alterungsverhalten mischleitender LSCF Kathoden für Hochtemperatur-
           Festoxid-Brennstoffzellen (SOFCs). 2011
           ISBN 978-3-86644-652-6

Band 20   Annika Utz
           The Electrochemical Oxidation of H_2 and CO at Patterned Ni Anodes of
           SOFCs. 2011
           ISBN 978-3-86644-686-1

Band 21   Jan Hayd
           Nanoskalige Kathoden für den Einsatz in Festelektrolyt-Brennstoffzellen
           bei abgesenkten Betriebstemperaturen. 2012
           ISBN 978-3-86644-838-4

Band 22   Michael Kornely
           Elektrische Charakterisierung und Modellierung von metallischen Inter-
           konnektoren (MIC) des SOFC-Stacks. 2012
           ISBN 978-3-86644-833-9

Band 23   Dino Klotz
           Characterization and Modeling of Electrochemical Energy Conversion
           Systems by Impedance Techniques. 2012
           ISBN 978-3-86644-903-9

Band 24   Alexander Kromp
           Model-based Interpretation of the Performance and Degradation of
           Reformate Fueled Solid Oxide Fuel Cells. 2013
           ISBN 978-3-7315-0006-3